HumblePi

The Role Mathematics *Should* Play in American Education

Michael K. Smith, Ph.D.

Prometheus Books

59 John Glenn Drive
Amherst, New York 14228-2197

Published 1994 by Prometheus Books

98 97 96 95 94   5 4 3 2 1

Library of Congress Cataloging-in-Publication Data

Smith, Michael K., Ph.D.
    Humble pi : the role mathematics should play in American
education / by Michael K. Smith.
        p.   cm.
    Includes bibliographical references.
    ISBN 0-87975-877-5
    1. Mathematics—Study and teaching—United States. I. Title.
QA13.S65   1994
510'.71'073—dc20   94-3085
                                                         CIP

Printed in the United States of America on acid-free paper.

Contents

Foreword

When everyone around us seems to be lamenting the fact that math education is not what they think it should be, and manufacturing suggestions about how to improve it has become a minor industry, it is refreshing to read a book that challenges us to ask why we think it so important anyway. Michael Smith's examination of the reasons that drive the math education community leads to a categorical answer: the emperor has no clothes. *Humble Pi* is not a tirade against mathematics as such. Nor does it want to deny the usefulness (even in the age of calculators) of being able to figure that two dimes and a nickel make a quarter. What it does with wit and energy is punch holes in the reasons given for insisting that everyone should learn to do all that other stuff in the math curriculum. Do you really need to know how to divide seven-eighths by three-quarters or solve quadratic equations? Do you really believe that all those hours, pleasant for a few, painful for most, of repetitive math exercises taught you to think better?

People who write truly iconoclastic books cannot expect unqualified agreement even from readers sufficiently warmly disposed to write a foreword. But it is of no importance that the emperor might have a few rags after all. Accepting a fraction (if I dare use a mathematically tainted word) of Michael Smith's arguments will be enough to leave you asking how such a large-scale business as math ed. could be based on such flimsy reason to believe in the value of any possible return. The costs of math ed. are serious: in dollar terms, several hundred billion a year across the world; in psychological terms, tens of millions of egos damaged by a sense of failure; and much more as well. You would think that nobody would incur this kind of cost without the most solid reasons to believe in the benefits—and indeed such hard-nosed cost-benefit analysis is demanded in support of proposals for new educational plans. But this bias in burden of proof is a major factor making for conservatism in education (and not only in math ed.): to adopt something new requires more expensive proof than responsible innovators can afford; in the meantime we continue doing what we always did just because it is there even in the absence of any proof whatsoever of its value. In business this would be a formula for rapid bankruptcy. In education it appears that nations can continue mindless policies for longer periods before hitting disaster. It remains to be seen how long.

A semantic distinction will help avoid a mistaken impression that the conclusions of *Humble Pi* are in conflict with the positions I have adopted in my own writing on the future of mathematics in the intellectual lives of children. I use the word "math" to refer to what is taught in school and consider this as no more entitled to be called "mathematics" than I think the silly jingles you might hear in a TV commercial are entitled to be called "poetry." Now, in this language on almost every occasion where *Humble Pi* uses the word "mathematics" in talking about school stuff I would use the word "math"—and if this translation is made, I agree fully with the main theses of the book. Where I would make an addition is that I am optimistic about the future of new forms of true mathematics for children. They will look nothing like math. They may not even be recognized as such by most teachers. But they will come, and the more people read *Humble Pi* the quicker this will happen.

Seymour Papert
Massachusetts Institute of Technology

Acknowledgments

I would like to thank Ohmer Milton for his comments, criticisms, and intellectual support on this project, from the initial publication of what is now chapter 2, "Why Is Pythagoras Following Me?" in *Phi Delta Kappan*, through the development of this entire book. The editors at *Phi Delta Kappan*, especially Bruce Smith, offered generous advice on the original article and subsequent chapters. Richard Wisniewski, Dean of the University of Tennessee's College of Education, has nurtured a creative atmosphere that allowed me to undertake the critical questioning of assumptions that I do in this book. Steven Mitchell, my editor at Prometheus Books, provided detailed suggestions for revisions. Finally, Katie Rea has provided encouragement and support, and her excellent editing skills, throughout the entire process of completing this book.

Introduction

Humble Pi addresses the myriad issues that are linked by a belief in the supremacy of mathematics. The origins of the book were simple. In speaking to students, educators, engineers, and mathematicians, I thought I had asked a rather straightforward question: Why do we require our students to learn so much mathematics? In the flurry of educational reform movements of the eighties and nineties, however, when all indices seemed to indicate that the performance of U.S. students was at its lowest in two decades, in fact that it was the worst in the world, the answer to this question seemed obvious: because math is so important. To suggest otherwise seemed almost blasphemous.

My question, however, provoked responses, even from the mathematicians, that were more emotional than logical. I began to feel like I was asking Why Christmas? Why the World Series? If the emotions I had seen were all negative, in the sense of being against my views, then it would have been natural to doubt myself..However, the emotions were strongly Janus-faced: in opening my mail, one letter would praise me, then another letter

11

would condemn me to purgatory. On the laudatory side, many readers thanked me for questioning some of the sacrosanct notions surrounding mathematics and for noting particularly the terrible anxiety some people have in learning a subject that they consider useless. Other readers were more derogatory critics: how could I possibly think that the study of any mathematics was unnecessary, especially when the United States seems to be lagging so far behind other countries (i.e., Japan) in scientific and technological prowess?

It seems to me, though, that every once in a while we should stop and question our cherished notions. Perhaps the questioning will reaffirm our beliefs; perhaps we will have to rethink our answers. In any case, I had always thought, and was even taught, that discussion and questioning form the basis of critical thinking, an ability that educators seem to care about and want their students to possess. In this vein, then, take the present book as some critical thinking about mathematics and mathematics education.

This book is intended to challenge some of our cherished notions about the role of mathematics and mathematics education in our present society. I am not anti-math, but I do believe there are certain myths and assumptions which seem to be reified when we start talking about mathematics. Let my questions serve as guides to question these assumptions.

For instance, is the ability to reason mathematically linked to the ability to reason logically? Are those people who are mathematically sophisticated better able to discern logical flaws in arguments? John Allen Paulos, in his book *Innumeracy*, answers yes to both questions. The first chapter, "Humble Pi," challenges Paulos's belief that simple numeracy, i.e., the ability to solve rather abstract mathematical exercises, is linked to logical reasoning ability.

Why do we want students to take so many mathematics courses, especially at the high school level? Do most people need courses in algebra, geometry, and trigonometry? The second chapter, "Why Is Pythagoras Following Me?", addresses these questions by examining three virtually unchallenged assumptions that lie at the heart of mathematics education. First, it is assumed that the study of mathematics is necessary to understand the structure of the world, and even of society, since both can be described so eloquently by mathematical formulae. Is this true? Is mathematics so predictive?

Second, many mathematicians and mathematics educators claim that the study of mathematics prepares a student for other endeavors, makes them more logical and concise, and better critical thinkers. Thus, to do badly in mathematics makes one illogical, confused, and stupid. This is Paulos's argument as it exists in educational circles. In many cases, the belief in this assumption affects a student's future: bad grades and test scores hinder progress toward high school and college graduation, make entrance to certain colleges and graduate programs difficult, and make a student less competitive for scholarships. I'm thinking particularly of the great weight given to admissions exams, such as the Preliminary Scholastic Assessment Test (PSAT),* the Scholastic Assessment Test (SAT),† and the Graduate Record Exam (GRE),‡ all of which base one-half of their total score on

---

*Primarily taken by high school sophomores and juniors who are competing for National Merit College Scholarships.

†Primarily taken prior to applying for entrance into undergraduate programs. The SAT has recently changed its name from Scholastic Aptitude Test to the Scholastic Assessment Test.

‡Primarily taken prior to applying for entrance into graduate schools.

mathematics performance. Are we illogical if we can't master mathematics? What is the evidence on which this claim is based?

Third, mathematicians and mathematics educators claim that the study of mathematics is necessary for daily living. You need these mathematics and will use them for many practical purposes, they say. Thus, a study of algebra, geometry, trigonometry, and calculus is essential for success in life, not only in scientific and technological careers but in other pursuits as well. How much math do we really need to become educated persons, and well-rounded citizens?

The widespread belief that a study of mathematics will help us understand the world, make us more logical thinkers, and be useful in everyday life permeates all discussions of mathematics education. It is these assumptions that need to be challenged. It is also the consequences of these assumptions that we need to examine closely if we are truly to understand the mathematical noose that is strangling current educational discussions.

The remaining chapters discuss various consequences that arise from a belief in the supremacy of mathematics.

Chapter 3, "The Word Problem," examines typical mathematics problems that occur on standardized exams, such as the SAT. Admissions to colleges and universities are determined by scores on tests like the SAT and by performances on these "word problems." Test makers feel that they can measure a student's general intellectual and logical abilities by these types of problems. If this is true, why do most of these problems seem so absurd?

Chapter 4, "Testosterone Trigonometry," examines the belief that males are superior to females in mathematical ability and that this difference is biologically determined. This male superiority in mathematics also implies a greater skill in logical reasoning abilities. How do we know that males are better at mathematics than females? How strong is evidence for inheritability of this ability? What other factors might contribute to someone excelling in mathematical ability?

Chapter 5, "Last Place Blues," challenges the notion that the United States is in last place in the world in mathematical ability and that our last place position is due to a deteriorating educational system. It also challenges President Clinton's suggestion that, by the year 2000, our students should be number 1 in the world in mathematics and science, based on international tests. Is our educational system so terrible? What would it really mean to lead the world in mathematics and science?

Chapter 6, "Let No One Enter Who Does Not Know Geometry," considers how mathematics has been taught over the last hundred years and outlines the argument among mathematics educators over the teaching of pure mathematics versus practical mathematics. It also asks the question, How well do mathematics concepts that we learn in school transfer to more realistic situations?

Chapter 7, "How Much Mathematics Is Really Needed?", challenges the ideas of a national curriculum and a national test in mathematics. It also argues that separate courses in mathematics, especially at the high school level, should be eliminated. Instead of traditional mathematics courses, students should work on solving complex problems, problems which draw on knowledge from different disciplines and which are more closely connected to experiences they will face as citizens.

1

Humble Pi

We live in a world dominated by mathematicians. We are constantly told how important mathematics is for success in the world, how much it contributes to our country's economic strength, and how essential it is to the education of our children. We are also constantly reminded how dumb most of us are when it comes to doing mathematics. Not only are we "innumerate"[1] but our country's educational structure has deteriorated so much that our students are now in last place in the world in mathematical ability.

Am I exaggerating? Just listen to one national report:

The mathematical skills of our nation's children are generally insufficient to cope with either on-the-job demands for problem solving or college expectations for mathematical literacy. Because of the emergence of the importance of mathematics to so many areas of education, citizenship, and careers, business and industry spend billions in training, colleges and universities devote large amounts of resources to remediation, and still the

United States is having difficulty maintaining its competitive edge in the global marketplace.[2]

Or another report:

Looking toward the year 2000, the fastest-growing occupations require employees to have much higher math, language, and reasoning capabilities than do current occupations. . . . [But too] many students leave high school without the mathematical understanding that will allow them to participate fully as workers and citizens in contemporary society.[3]

Or one final report:

Mathematical and scientific literacy has become even more crucial to developing a citizenry capable of making informed judgments. As public policy issues increasingly involve scientific and technological issues, "the preservation of democratic governments in the 21st Century may depend on the expansion of the public understanding of science and technology."[4]

In a single breath, our ability to do mathematics is linked with getting a successful job, making informed judgments on social issues, being good citizens in a democratic society, and perhaps even with the survival of our democracy itself. If this picture is truly accurate, then we must seriously worry about what kinds of mathematics our children learn in school, and whether other subjects should be excluded to prepare them fully in mathematical skills.

Is mathematics as important as these studies claim? What mathematics do our students learn and what kinds of mathematics tests are they given? On what basis are judgments made about the importance of mathematics? How do we know that mathematics is so important to being a good citizen? How do we know that mathematics contributes so much to our economic success as a nation?

In considering the importance of mathematics, at least three factors stand out. The first relates to *mathematical calculation:* exactly how much arithmetic, algebra, geometry, trigonometry, calculus, or other mathematics does the average person need to perform a particular job or to make informed judgments about societal issues? The second relates to *mathematical integration:* how important is mathematics to occupations other than the academic field of mathematics itself? How much mathematics is necessary for an understanding of science, economics, history, or literature? Detailed discussions of these two questions will occupy subsequent chapters.

If these were the only two factors that related to the importance of mathematics, then our job would probably be much more straightforward. There is a third factor, however, that elevates questions about the importance of mathematics to a different level. This factor posits a *relationship between mathematics and logic:* Does the study of mathematics make us more logical? Does the study of the logic of mathematics readily transfer to an understanding of the logic of other disciplines? I believe that this assumption underlies many of the arguments for the supposed primacy of mathematics. Like the late nineteenth-century belief that no one could be considered educated or even logical without a knowledge of the intricacies of the Latin language, so today we seem to believe that mathematics is the key to intellectual and logical superiority. It is almost an arrogant belief in the power of mathematics to solve problems and to train the mind. Because of its fundamental nature, we must first address this assumption before tackling issues related to the practicality of mathematics.

Mathematics and Logic

> What shall we say, Simplicio? Must we not confess that geom-
> etry is the most powerful of all instruments for sharpening the
> wit and training the mind to think correctly? Was not Plato
> perfectly right when he wished his pupils should be first of
> all well grounded in mathematics?
> —Galileo, *Dialogues Concerning The Two New Sciences*[5]

One of the more prevalent assumptions about why mathematics
is important to an educated person is the belief that the ability
to reason *mathematically* is strongly linked to the ability to reason
logically. In other words, people who learn mathematical methods
are in a sense more logical, more informed, and even more thoughtful. The inverse of this proposition is humbling: If you cannot
reason mathematically, then you are not a very logical person.

John Allen Paulos argues this assumption very strongly. In
his best-selling book, *Innumeracy*, Paulos tries to show the lack
of mathematical and logical "common sense" in the American
public. Paulos's grand theme is that the ability to be "numerate"
and the ability to be logical are intimately interconnected. There
is no better place to start testing this assumption than in examining the evidence in Paulos's book.

Paulos is distressed that most of the country is suffering
from "innumeracy," which he claims is an inability to deal com-
fortably with the fundamental notions of number and chance.
"I'm always amazed and depressed," he says, "when I encounter
students who have no idea what the population of the United
States is, or the approximate distance from coast to coast, or
roughly what percentage of the world is Chinese."[6] Why should
he be so distressed? If people were capable of simple estimation
and calculation, Paulos contends, then fewer ridiculous notions

would be entertained. Perhaps we wouldn't still believe Elvis is alive or that Martians got Lucille Ball pregnant!

Paulos is drawing on a long tradition in Western culture that links mathematics and logic. The philosopher René Descartes, in his *Rules for the Direction of Mind*, states as his primary assumption, "We reject all such merely probable knowledge and make it a rule to trust only what is completely known and incapable of being doubted." This leads to the importance of mathematics:

> But one conclusion now emerges out of these considerations, viz. not, indeed, that Arithmetic and Geometry are the sole sciences to be studied, but only that in our search for the direct road towards truth we should busy ourselves with no object about which we cannot attain a certitude equal to that of the demonstrations of Arithmetic and Geometry.[7]

Francis Bacon sees mathematics as a tool for training the mind, in his *Advancement of Learning*:

> In the mathematics I can report no deficiencies, except it be that men do not sufficiently understand the excellent use of pure mathematics, in that they do remedy and cure many defects in the wit and faculties intellectual. For if the wit be too dull, they sharpen it; if too wandering, they fix it; if too inherent in the sense, they abstract it.[8]

The interconnection of mathematics and logic often finds its way into literature. In Leo Tolstoy's *War and Peace*, the old Prince Nicholas Andreevich Bolkonski outlined his daughter's (the Princess Mary) education as follows:

He used to say that there are only two sources of human vice—
idleness and superstition, and only two virtues—activity and
intelligence. He himself undertook his daughter's education, and
to develop these two cardinal virtues in her gave her lessons
in algebra and geometry till she was twenty, and arranged her
life so that her whole time was occupied.

During this tutoring, the Princess Mary at one point gave a wrong
answer to a geometry problem. The Prince scolded her: "Mathematics are most important, madam! I don't want to have you
like our silly ladies. Get used to it and you'll like it," and he
patted her cheek. "It will drive all the nonsense out of your head."[9]

Paulos is certainly in this tradition of authors who claim
a vital linkage between mathematics and logic. For instance,
deductive logic operates by establishing a set of premises and
then drawing valid conclusions based on these premises. The
processes of mathematical reasoning are seen as operating in
a similar fashion. Euclid established several basic premises which
he assumed to be true of space and time; from these he deduced
elements and propositions of his geometic theory, which he could
prove followed logically from his initial premises. It was only
natural that philosophers would link logic and math to life and
claim that an understanding of the former would help one deduce
a logical course in the latter.

Paulos is upset that so many people today are illogical because
they are innumerate and lack the ability to solve simple mathematical problems. But are his problems so simple? Let's try to
work through a few of his examples.

To see what Paulos means by *simple* let's consider one of
his examples: what is the volume of all the human blood in the

world?* Give up? According to Paulos, the average adult male has about six quarts of blood. There are about 5 billion people in the world, so they contain about 5 billion gallons (5×10^9) of blood. Since there are about 7.5 gallons per cubic foot, this comes to approximately 6.7×10^8 cubic feet of blood. (Just in case you want more math, the cube root of this last number is 870; therefore all the blood in the world would fit into a cube 870 feet on a side.)

Try another problem: In how many ways could eight leaders of the Western countries be arranged for a group photo? When the leaders of eight Western countries get together, there are $8 \times 7 \times 6 \times 5 \times 4 \times 3 \times 2 \times 1 = 40{,}320$ ways to arrange them for a group photo. Now, this is one exercise that I would actually like to try: Ok, Clinton next to Rabin, now next to Major, now at the end of the line. Shape up, Bill.

Or one final problem: What are the chances you just inhaled a molecule that Julius Caesar exhaled in his dying breath? Paulos assumes that the exhaled molecules are still floating around in the atmosphere after two thousand years. Let N stand for the number of molecules in the world, A for the number exhaled by Julius Caesar in his dying breath; then the probability that you have inhaled a molecule from Caesar is A/N. The probability that you have not inhaled a molecule from Julius Caesar is 1-A/N. If you inhaled three molecules, then the probability of none of these being from Caesar is $[1 - A/N]^3$. The probability of the complementary event, that you have inhaled a molecule from Caesar, then becomes $1 - [1 - A/N]^3$. Since A is fairly small,

---

*Please no comments yet about why anyone would do this problem. Just work the problem: as Paulos says, "These estimations are generally quite easy and often suggestive."[10] Readers are urged to try these problems and judge for themselves how simple they are.

and N is huge, this probability becomes about .99, indicating that there is a good chance you have just inhaled a molecule breathed by Julius Caesar in his dying breath. It's that simple!

As the reader can see, Paulos's problems are not quite as simple as knowing how far it is from New York to Los Angeles or estimating the population of the United States. To Paulos, problems like estimating the volume of human blood in the world are "simple" mathematical exercises. Furthermore, not knowing how to do simple calculation and estimation is also indicative of not being a logical person. "Since numbers and logic are inextricably intertwined both theoretically and in the popular mind, it's perhaps not stretching matters too far to describe faulty logic as a kind of innumeracy."[11]

This faulty logic is rampant, as evidenced by such "illogical" popular beliefs in "pseudosciences," such as parapsychology, astrology, numerology, and the belief in UFOs. But innumeracy is not confined to the layperson; it is pervasive in academics as well. Paulos then suggests that certain theories, such as those advocated by Karl Marx and Sigmund Freud, are not only unscientific but innumerate and illogical as well. After suggesting that Freud was innumerate, he states, "This is certainly not the place to argue whether or not Freudianism and Marxism should be deemed pseudosciences, but a tendency to confuse factual statements with empty logical formulations leads to sloppy thought."[12]

Mathematics, combined with the scientific method, allows us to distinguish between absurd claims and hard evidence.

Less controversial is the contention that there are no clearcut, easy algorithms* that allow us to distinguish science from pseudoscience in all cases. The boundary between them is too

---

*A set of rules for solving a problem in a finite number of steps.

fuzzy. Our unifying topics, number and probability, do, however, provide the basis for statistics, which, together with logic, constitutes the foundation of the scientific method, which will eventually sort matters out if anything can.[13]

The study of mathematics can not only help us distinguish hard science from pseudoscience, but also change our lives.

> Women, in particular, may end up in lower-paying fields because they do everything in their power to avoid a chemistry or an economics course with mathematics or statistics prerequisites. I've seen too many bright women go into sociology and too many dull men go into business, the only difference between them being that the men managed to scrape through a couple of college math courses.[14]

I imagine the old Prince Bolkonski today would have urged his daughter to become an MBA instead of a sociologist!

Furthermore, the study of mathematics can almost assure us of a decent job:

> The students who do major in mathematics in college, taking the basic courses in differential equations, advanced calculus, abstract algebra, linear algebra, topology, logic, probability and statistics, real and complex analysis, etc., have a large number of options, not only in mathematics and computer science but in an increasing variety of fields which utilize mathematics. Even when companies recruit for jobs that have nothing to do with mathematics, they often encourage math majors to apply, since they know that analytical skills [i.e., logical reasoning skills] will serve anyone well, whatever the job.[15]

Mathematics, then, is a powerful tool—it can combat innumeracy, the pseudosciences, and the illogical. It can help people get better jobs or change their lives. Even if mathematics could do none of these things, it still has the power to help us feel more knowledgeable than other people. Paulos recounts the story of an argument he had with one of his teachers about the correct way to compute an Earned Run Average (ERA) for a baseball pitcher. When his calculation turned out to be the correct one, he notes: "I remember thinking of mathematics as a kind of omnipotent protector. You could prove things to people and they would have to believe you whether they liked you or not."[16]

Whether or not the study of mathematics can change lives or jobs is something that will be considered later. I can sympathize, however, with Paulos when he talks about the power of mathematics. I can remember using mathematics as a youngster to achieve a sense of power, a feeling that I knew more than my friends and my parents. I loved mathematical tricks. In the eighth grade, I joined with four friends to form a group in which we shared ideas. Someone would learn something, and then come and teach the group. One day I posed a challenge: I bet I can add up the numbers from 1 to 100 *in my head* faster than all four of you can do it together with pencil and paper. They took my challenge and began calculating as fast as they could. Little did they know that I had read about a trick (probably in one of Martin Gardner's many excellent books on mathematics) that showed how to add up consecutive numbers quickly. In fact, I knew all along that the answer was 5050. After my friends had sweated for several minutes, I announced the answer. Of course, they didn't believe me until they had finished their own calculations. (Here's how it works: the formula for adding up numbers from 1 to n is $n(n+1)/2$. So for the numbers from 1 to 100, this becomes $100(100 + 1)/2$ or 50 times 101 which is 5050.)

There was a trick involved in this example. I never thought otherwise. I loved discovering these mathematical tricks and shortcuts. I was delighted as I continued, throughout high school, to discover that these mathematical tricks often occurred on standardized exams. I remember one particular problem from my SAT (Scholastic Aptitude Test as it was called then) when I was a junior in high school: How much greater is the sum of the numbers from 101 to 200 than the sum of the numbers from 1 to 100? It didn't take me too long to answer 15,050. (Use the same formula as above. The sum of the numbers from 1 to 200 is 200(200 + 1)/2 or 100 times 201 which equals 20,100. Then subtract the sum from 1 to 100, which is 5050, to get the answer of 15,050.) But by this time I had plenty of practice with these types of problems.

The trouble with Paulos's view is that he wants us to believe that the ability to solve these types of problems involves something more than mathematical tricks. He claims that the ability to solve such numerical puzzles somehow involves a higher degree of logical thinking or "analytical skills." The conclusion then becomes obvious: Only logical, numerate people can solve these problems; an inability to solve them indicates innumeracy or an illogical frame of mind.

I'm not convinced. Let me give some more examples of why I'm skeptical, and why you should be, too. In the following list are presented six examples of problems that Paulos claims a numerate person should be able to answer, given a little thought and some calculation. Try them yourself.

1. How fast does human hair grow in miles per hour?

2. How many people die on earth each day?

3. How many cigarettes are smoked annually in this country?

4. How many people would there have to be in a group in order for the probability to be half that at least two people in it have the same birthday?

5. What are the chances that you just inhaled a molecule which Julius Caesar exhaled in his dying breath?

6. Assume that Myrtle has reason to believe that she'll meet N potential spouses during her "dating" life. She asks herself the question: When should I accept Mr. X and forgo the suitors who could come after him, some of whom may possibly be "better" than he?[17]

Paulos solves all of these problems quickly and offers very definite numerical answers for each one (as we have already seen for problem number 5, the Julius Caesar example). I was a little more skeptical of his quick solutions. For instance, the first three problems suggest that a quick trip to the library wouldn't hurt: Off the top of my head, I don't recall average death rates for certain countries or what percentage of the populace smokes. From having taught probability theory, I know that problem four is nasty. Problem five seems really far-fetched. Finally, with any reasonable answer to number six, Paulos should start writing self-help books. He'd make a fortune!

Perhaps, however, I'm not the best judge of how quickly these problems can be solved by a trained mathematical mind. So I sent these math problems to the entire math faculty at a major university. I asked the mathematicians for their help in solving these problems. I told them the problems came from Paulos's book and that he suggested they had definite numerical answers. I asked them to state any assumptions that needed to be made and to describe any calculations they had used in arriving at their answers.

On the tests that were returned to me, not a single one of

Paulos's problems was worked. However, there were numerous comments, or should I say complaints, about the problems: "There is no definite answer to this problem unless you refine the questions considerably"; "You would need to make an enormous number of assumptions to answer these questions"; "There is no answer to this without making lots of assumptions"; "None of the above are well specified enough to answer"; and, finally, "These are not really 'math' problems but problems of estimation. Mathematicians usually don't bother with these types of questions!" Furthermore, very few mathematicians responded to this test: is this because they're too busy, too lazy, too uninterested, or themselves too innumerate?!

This view of Paulos's work is not unique to my sample of mathematicians. Lisa J. Evered, reviewing the book in *The American Mathematical Monthly*, asks, "Who is interested in the kinds of numerical calculations the author exemplifies?"[18] She continues:

Archimedes calculated the number of grains of sand it would take to fill the universe because he wanted to show that such a calculation could be done. Granted the Greek system of numeration, it was no mean feat. But apart from the transient sense of triumph, did anyone really care what number he came up with? The same thing is even more obviously true with Professor Paulos's own example: the depth of all of the human blood in the world if it were stored in a reservoir in Central Park? Many numerates would have to agree with the innumerates: Who cares? Nonetheless, some of us definitely enjoy working such things out. But the key phrase is "some of us." In fact, it is very few of us, and most mathematicians are not included. My observation is that the enjoyment of such calculations is not a common characteristic, even among numerate people. The reason for the enjoyment seems to be partly that of the mountaineer: It is there![19]

To claim that the inability to solve these types of problems makes one innumerate is ridiculous. Evered makes an analogy with illiteracy. Functional illiteracy is the inability to read and to handle the most basic functions that reading entails, such as reading signs or following written instructions. "However, we surely do not argue that anyone has some degree of illiteracy that is of social concern simply because that person does not ever read Shakespeare or Thomas Mann."[20] The choice of specific authors is a matter of certain tastes in literature, rather than a basic problem with literacy.

Paulos's examples are of the same type. His problems involve beliefs and habits and interests, the games that people like to play with numbers. He suggests that we should know the elements of probability theory, as in his problem of the birthdays. As Evered suggests, the teaching of probability theory is trying, even for highly intelligent students. "One reason for this," she says, "is undoubtedly that the methods of solution of elementary probability problems often seem to the student to smack of ad hoc conjuring tricks."[21] The reader should reread the Julius Caesar example for evidence of what Evered is alluding to.

Evered concludes by arguing for a definition of functional innumeracy:

> This would surely mean such things as inability to calculate an 8.25% sales tax on a purchase, inability to calculate the price that results from a 20% reduction in price, inability to fill out income tax forms, inability to calculate the paint required for an apartment given the square feet covered per gallon, and so forth. These are the kinds of things that everyone needs to be able to do—and can learn to do.[22]

In both chapters two and seven, I'll have more to say about the types of mathematics the average person should know.

We have seen, in these few examples, some of the problems in doing "simple" mathematics. We have also seen that even mathematicians may not agree on the importance of the problems that Paulos claims makes one a "numerate" person. But people who feel they are numerate, or who like to play mathematical games, often aren't nearly so tolerant of people who can't do "simple" mathematics. "Mathematics is logic," we can almost hear them say, "and if I can do problems you can't, then this proves how much smarter I am than you." Let's look at another example.

Ask Marilyn

Readers of *Parade* magazine in their Sunday newspaper have probably noticed a regular column entitled "Ask Marilyn," written by Marilyn Vos Savant. According to the *Guinness Book of World Records Hall of Fame,* Marilyn is listed as having the world's highest IQ. In her column, readers try to befuddle her with all kinds of mathematical, logical, and other erudite puzzles.

One reader posed the following challenge:

> Suppose you're on a game show, and you're given the choice of three doors: Behind one door is a car; behind the others, goats. You pick a door, say No. 1, and the host, who knows what's behind the doors, opens another door, say No. 3, which has a goat. He then says to you, "Do you want to pick door No. 2?" Is it to your advantage to switch your choice?

Marilyn answered it this way: "Yes; you should switch. The first door has a one-third chance of winning, but the second door has a two-thirds chance."[23]

I must admit that the Sunday morning I first read this puzzle I thought Marilyn had made a mistake. It seemed to me that, if there were two doors left, then the odds should be even (1/2 to 1/2) as to which door had the car. Her suggestion that the best odds were provided by switching seemed to be counter-intuitive. At this point, I dismissed the argument, not that I thought I was perfectly right and that she was wrong, but merely that I had to have time to read the funnies that morning.

Most other readers were not so forgiving. In fact, over the next few weeks, scores of numerate people, mathematicians, and scientists crucified Marilyn for her mistake. Let me provide a sampling of the comments that were reprinted in *Parade*:

> I'll come straight to the point. In the [preceding] question and answer, you blew it! Let me explain: If one door is shown to be a loser, that information changes the probability to 1/2. As a professional mathematician, I'm very concerned with the general public's lack of mathematical skills. Please help by confessing your error and, in the future, being more careful.

> You blew it, and you blew it big! I'll explain: After the host reveals a goat, you now have a one-in-two chance of being correct. Whether you change your answer or not, the odds are the same. There is enough mathematical illiteracy in this country, and we don't need the world's highest IQ propagating more. Shame![24]

> You are in error—and you have ignored good counsel—but Albert Einstein earned a dearer place in the hearts of the people after he admitted his errors.

May I suggest that you obtain and refer to a standard textbook on probability before you try to answer a question of this type again?

Your logic is in error, and I am sure you will receive many letters on this topic from high school and college students. Perhaps you should keep a few addresses for help with future columns.

You are utterly incorrect about the game-show question, and I hope this controversy will call some public attention to the serious national crisis in mathematical education. If you can admit your error you will have contributed constructively toward the solution of a deplorable situation. How many irate mathematicians are needed to get you to change your mind?

You are the goat.

You're wrong, but look at the positive side. If all those Ph.D.s were wrong, the country would be in very serious trouble.

Maybe women look at math problems differently than men.[25]

It obviously doesn't pay to do mathematical problems incorrectly. The most intelligent woman in the world is excoriated for her lack of intelligence, for her inability to do mathematics, and even for her gender. (We will examine the belief that "women look at math problems differently than men" in a later chapter). The furor over this blatant public display of innumeracy might be justified except for one small point: I believe that Marilyn is right. Let's go back and look at this problem more closely.

In the original problem there are three doors, with goats behind two of the doors and a car behind the other one. To calculate probabilities, it is important to consider the possibilities of a given problem. There are three possible situations that could occur in this example:

Table 1

	Door 1	Door 2	Door 3
Arrangement 1	Goat	Goat	Car
Arrangement 2	Goat	Car	Goat
Arrangement 3	Car	Goat	Goat

These are the logical possibilities for how we could arrange the goats and the cars. Now, you as the contestant are allowed to pick one door. Using the table above, it should be clear that if you pick Door 1, then you have one chance out of three of winning the car (i.e., under Door 1, twice you get goats and only once do you get a car). If you pick Door 2 or Door 3, the same result obtains: only one time in three would you win the car. So in this statement of the problem, the contestant only has a one-third chance of winning the car no matter which door is picked.

Now, the problem becomes more complicated. The host of the show knows what's behind every door. Whenever you, the contestant, pick a door, the host is going to reveal what's behind another door according to the following rule: the host always reveals a door that has a goat and not a car. Let's say you pick Door 1. Our diagram now becomes as follows:

Table 2

	Door 1	Door 2	Door 3	Switch	Stay
Arrangement 1	Goat (your pick)	Goat (Host reveals this door)	Car	Win	Lose
Arrangement 2	Goat (your pick)	Car	Goat (Host reveals this door)	Win	Lose
Arrangement 3	Car (your pick)	Goat	Goat (Host reveals this door)	Lose	Win

You as the contestant are picking Door 1 and the host is always revealing a door behind which is a goat. Now, the important question is: Should you switch doors once the host reveals a door with a goat? The answer is a resounding yes. Look at the diagram. In the first arrangement, if you switched doors (i.e., gave up Door 1 and switched to Door 3), you win the car. In arrangement 2, if you switched doors, you also win the car. Only in arrangement 3 is switching not beneficial and you give up the car for the goat. So, in two arrangements out of three, you win the car by switching. In other words, if you switch you have a two-in-three chance of winning and a one-in-three chance of losing. Thus, the odds are in your favor to switch and Marilyn is correct.

How did I figure this out? Am I a smarter or more logical person than all those other Ph.D.s who chided Marilyn on her incompetence? Obviously, I would like to think so but my modesty suggests that I consider other factors. For instance, part of my duties as a university professor in psychology and education is to teach undergraduate and graduate courses on statistics and probability theory for the social sciences. As such, I have spent hours trying to understand the intricacies of certain aspects of this domain in order to be able to explain certain principles in a common-sense fashion. I have tried to comprehend all kinds of esoteric probability problems and statistical puzzles.

Even with all this training I missed Marilyn's problem the first time I tried to solve it. Only after I examined it in more detail and used some of the probability training that I had learned for my courses did I arrive at the answer above. And that answer may still not be entirely correct. Maybe I still missed something.

But who cares?

Lotteries, Luck, and Logic

State lotteries are slowly becoming ubiquitous. As states scramble for more revenue, the lure of lottery dollars, as opposed to more tangible forms of taxation, is becoming almost overwhelming. When I teach either undergraduate or graduate statistics courses, I caution against the lure of lotteries and dutifully calculate the odds against winning huge sums of money. It seemed foolish to me for anyone to play lotteries since the odds were so much against winning.

This was my thinking until one of my graduate students won the Kentucky lottery. Then I began to wonder why people gamble or take risks or why someone willingly goes against the odds. This particular student was from Malaysia and had the habit of occasionally playing lotteries in whatever country or state he currently resided. Now, Tennessee itself doesn't yet have a lottery system. So my student had to pay someone to cross over into Kentucky and purchase his ticket for him. After he discovered that he had won, he waited two weeks to claim his $100,000 prize. Why? In his own words, he was busy studying for midterm examinations and couldn't afford to take the time off! After he did have time, he took a bus to Louisville (about six hours) to collect his prize money. This story made the front page of our local newspaper.

Was it reasonable for him to play the lotteries? Is he an innumerate person because he risked a one-dollar ticket to win a huge prize? How are we to account for luck, or the unexpected, or the long shots in any discussion of how mathematics and logic relate?

Let's try and be mathematical about this matter, first. My student's winning number was 01 22 29 33 35. He played what is called the Kentucky Cash Five. In this game, a contestant is allowed to pick any five numbers between 1 and 35. During the daily drawings, 35 ping pong balls, each with one of the numbers 1 to 35, are placed in a type of hot-air container (something like the hot-air popcorn makers). When the person conducting the lottery drawing releases the top of the container, a ping pong ball randomly pops to the top. The first five such balls then become the winning numbers for the day. To win the grand prize, a contestant has to have the winning five numbers, but in any order. In other words, my student would win if the numbers emerged in the order 01 22 29 33 35 or 22 29 01 33 35 or any other possible ordering.

What are the odds against winning? There is a formula for combinations for this type of problem, but let's see if we can calculate the probabilities a little more intuitively. First, consider how many different ways someone could choose five lucky numbers. The first number could obviously be any of the 35 numbers; the second number could be any of the 34 remaining numbers; the third number any of 33 remaining numbers; and the fourth and fifth numbers, any of the 32 and 31 remaining other numbers. There are then 35 × 34 × 33 × 32 × 31 = 38,955, 840 different orderings of five numbers drawn from 35 numbers.

So far, however, this is a bit of an underestimate of our

chances of winning. What we have calculated so far are all the different orderings of five numbers. Thus, we have possibilities for combinations such as 02 19 22 33 34 and 05 17 21 22 30, which are clearly different from one another. But we also have the following types of orderings: 01 22 29 33 35 and 01 29 33 35 22 and 22 29 33 35 01. These three examples all contain the same five numbers, but in different orders. A contestant would win if any of these orderings emerged when the ping pong balls were drawn. In fact, there are $5 \times 4 \times 3 \times 2 \times 1 = 120$ different ways to order any five numbers. (Try this on a lazy Sunday afternoon.)

The true probability of winning then becomes our first number, 38,955,840 different orderings, divided by 120 possible arrangements of any five numbers. This yields 324,632 different combinations of any five numbers. Thus, my student had one chance in 324,632 of picking the one winning five-number combination.

Was it logical to play the lottery? Certainly I'm only betting one dollar for a chance to win $100,000. But logically you have only one chance in over 300,000 of winning. Someone could spend a lot of their annual salary (lots of one-dollar tickets) before you would even come close mathematically to winning.

Why would anyone then undertake such a gamble? I believe that some human beings feel they are exceptions to mathematical logic. They believe that they can go against the odds to win something, or do something, or discover something. I would argue that this belief of beating the odds spurs many such human activities, including experimenting with miracle cures for certain diseases, advocating unpopular or controversial notions, and even pursuing unlikely scientific hypotheses.

Certainly some such contemporary beliefs appear patently ridiculous. The field of psychology is filled with self-help therapies, all guaranteeing to make people feel better about themselves and their lives. For instance, recent New Age therapies for healing injuries and restoring a person's "psychic balance" include: *aromatherapy*—use of essential oils from plants and animals massaged into the skin; *crystal healing*—the derivation of healing energy from quartz and other minerals; and the *Alexander technique*— training to improve poor posture to alleviate pain. Frankly, I don't need to be a mathematician to have a healthy dose of skepticism about the efficacy of these cures.

Certain famous scientific discoveries, unfortunately, also appeared ridiculous in their own time. Einstein's Theory of Relativity took years to be widely accepted. Conversely, Einstein himself never could believe in the logic behind quantum mechanics, a theoretical system which now dominates certain aspects of physics. Darwin's theory of natural selection as tied to evolution wasn't wholeheartedly accepted in his time or in ours (Witness the continued attacks on Darwinian theory from "creationists" and even from certain scientific circles.) Edison persisted through thousands of fruitless experiments before he found the right filament that would make his idea of the electric light bulb work. Should we fault Edison for being so stubbornly persistent or should we admire him for struggling against the odds?

In his autobiography, Edison remarks on both the fortuitous circumstances surrounding discovery and the intense human effort needed to take advantage of such insights:

> In trying to perfect a thing, I sometimes run straight up against a granite wall a hundred feet high. If, after trying and trying and trying again, I can't get over it, I turn to something else. Then, someday, it may be months or it may be years later, something is discovered either by myself or someone else, or something happens in some part of the world, which I recognize

may help me to scale at least part of the wall. . . . I never allow myself to become discouraged under any circumstances. I recall that after we had conducted thousands of experiments on a certain project without solving the problem, one of my associates, after we had conducted the crowning experiment and it had proved a failure, expressed discouragement and disgust over our having failed "to find out anything." I cheerily assured him that we *had* learned something. For we had learned for a certainty that the thing couldn't be done that way, and that we would have to try some other way. We sometimes learn a lot from our failures if we have put into the effort the best thought and work we are capable of.[26]

This human endurance and persistence in something that is believed is what makes a simple mathematical or logical approach to the world unrealistic. Paulos recognizes the problem. It is very difficult to "disprove" certain claims or beliefs. He notes that the philosopher Willard Van Orman Quine argues:

Experience never forces one to reject any particular belief. [Quine] views science as an integrated web of interconnecting hypotheses, procedures, and formalism, and argues that any impact of the world on the web can be distributed in many different ways. If we're willing to make drastic enough changes in the rest of the web of our beliefs, the argument goes, we can hold on to our belief [in practically anything].[27]

This applies to miracle diets or cures for cancer or perhaps even beliefs in quantum mechanics.

The idea that "problems" can be "solved" by straightforward mathematics or logic is even doubtful when applied strictly to the practice of science. Does science "progress" by solving more and more complicated problems until it reaches a simple truth?

Thomas Kuhn, in his influential book *The Structure of Scientific Revolutions,*[28] has argued strongly against the belief that science can be seen as progressing closer and closer to the "truth" or some teleological goal.

Kuhn notes in detail the difficulties that physicists had, at the turn of the century, in accepting Einstein's view of the universe. For most physicists of the day, the Newtonian conception of the universe was the paradigm for understanding physical events and represented what Kuhn calls "normal science." Within normal science, the work of scientists is mostly seen as puzzle-solving, with criteria and techniques, whether logical or mathematical, guiding the everyday work of researchers.

Problems arise when anomalies, or unexpected facts, occur that cannot be "puzzled" away within an existing paradigm. The Michelson-Morley experiment was one such anomaly which presented problems that no Newtonian physicists could "solve." Originally designed to detect the presence of the "ether" which scientists felt permeated the universe and the fact that the speed of light would vary depending on the motion of this ether, the Michelson-Morley experiment instead demonstrated that the speed of light was constant, thus casting serious doubt on the widely accepted scientific concept of ether. These anomalies can provoke a crisis in normal science which could lead one of two ways: either the disturbing anomaly is ignored and put aside by scientists, or some new theory arises to explain the anomaly.

Scientific revolutions arise in these new theories that attempt to explain anomalies. Einstein was able to account for the results of the Michelson-Morley experiment while at the same time incorporating the major findings of Newtonian science. The only problem for most physicists was that Einstein's explanations posited an entirely different way of viewing the universe.

As Kuhn notes, many physicists resisted this new interpretation of the universe. Only as they began to get older, and a younger generation replaced them, did the Einsteinian view begin to become the new "normal science." Kuhn closes by describing the practice of science as almost Darwinian, analogous to the processes of natural selection. As he says:

> The resolution of revolutions is the selection by conflict within the scientific community of the fittest way to practice future science. . . . And the entire process may have occurred, as we now suppose biological evolution did, without benefit of a set goal, a permanent fixed scientific truth, of which each stage in the development of scientific knowledge is a better exemplar.[29]

This returns us once again to Paulos. Paulos desperately wants to believe in the omnipotence of mathematics, that it will solve our problems, that it alone can almost bring order to our physical and personal universes. "Our unifying topics, number and probability, do, however, provide the basis for statistics, which, together with logic, constitutes the foundation of the scientific method, which will eventually sort matters out if anything can."[30]

I'm not so optimistic. In fact, the world of real problems is much more complicated than the world of mathematical puzzles. Whereas the odds of winning lotteries, or the volume of blood in the world, or choosing which door to pick may have definitive (if somewhat tangled) solutions, more common problems—Is the greenhouse effect really a problem? When does life begin?—are amazingly complex human challenges. Humans may draw on mathematics or logic to help in the solutions of these problems, or their solutions may lie in a totally revolutionary point of view.

In fact, the major problem might reside in believing that there is a single "logic" that applies to all situations. Mathematicians certainly cannot be blamed for looking for exact, intrinsically beautiful solutions to difficult problems. But does their logic, their method of reasoning, fit other domains and other problems? Perhaps other situations are not so beautifully resolved with an exact solution, or perhaps the logic necessary in other endeavors is not parallel to that needed in mathematics.

Maybe there are several "logics," or at least different ways of conceptualizing what counts for evidence and for solutions in different domains. Howard Gardner, in his book *Frames of Mind: The Theory of Multiple Intelligences,*[31] makes a strong argument that there is not a single concept of "intelligence" but that there are multiple intelligences, each with its own structure. Gardner conceptualizes these intelligences as relatively autonomous "sets of know-how—procedures for doing things."[32] The mistaken belief in psychology is that there is a single intelligence, measurable by a single intelligence test.

Gardner's arguments, I believe, support my contention that there is no single logical way to approach every problem. In fact, different problems may call upon different intelligences. For instance, Gardner identifies the following six autonomous intelligences: linguistic, musical, logical-mathematical, spatial, bodily-kinesthetic, and personal. Each intelligence is structured according to its own set of procedures, its own "frame of mind."

One of Gardner's intelligences deals with logical-mathematical ability. His description of its procedures and its appeal match the enthusiasm that Paulos feels for the connection between mathematics and logic. There has certainly developed a strong bond between mathematics and logic, especially in this century. Bertrand Russell makes the strongest statement on the connection between the two disciplines: "Logic is the youth of mathematics and mathematics is the manhood of logic."[33]

As Gardner notes, however, the abilities in this domain don't easily transfer to abilities in other domains. Solving problems in mathematics or logic is not the same thing as solving problems in finance or the law. In fact, as Gardner suggests, mathematicians are seldom talented in other domains. Gardner quotes G. H. Hardy:

> It is undeniable that a gift for mathematics is one of the most specialized talents and that mathematicians as a class are not particularly distinguished for general ability or versatility. . . . If a man is in any sense a real mathematician, then it is a hundred to one that his mathematics will be far better than anything else he can do and . . . he would be silly if he surrendered any decent opportunity of exercising his one talent in order to do any undistinguished work in other fields.[34]

But is not the logic of mathematics the one true logic, or at least the most superior type of logic? Gardner struggles with this question:

> The drift of our society, and perhaps of other societies as well, raises sharply the question whether logical-mathematical intelligence may not be in some way more basic than the other intelligences. . . . It is often said: there is, after all, only one logic, and only those with developed logical-mathematical intelligences can exercise it.[35]

Gardner, however, finally has to disagree with this conception of logical-mathematical ability as the primary intelligence:

> To my way of thinking, it is far more plausible to think of logical-mathematical skill as one among a set of intelligences— a skill powerfully equipped to handle certain kinds of problems, but one in no sense superior to, or in danger of overwhelming,

the others. . . . There is indeed a logic to language and a logic to music; but these logics operate according to their own rules, and even the strongest dosage of mathematical logic into these areas will not change the ways in which their endogenous logics work.[36]

Certainly there are interactions between logical-mathematical intelligence and other disciplines, such as chess, engineering, and architecture. But there are limits:

It is possible to be a gifted sculptor, poet, or musician without having any particular interest in, or knowledge about, that orderliness and system that form the centerpiece of logical-mathematical thinking.[37]

Demystifying Mathematics

There is a certain mysteriousness that attaches to doing mathematics. People who can solve problems we can't, or who can solve problems more quickly than we might, seem to be more intelligent and swifter of mind. So far I have tried to show that there may be nothing magical about this mysteriousness. Many of the problems that a "numerate" person is supposed to be able to solve may indeed just be mathematical tricks and conundrums. In fact, many mathematicians dismiss the problems that Paulos says are important.

Mathematical problems are supposed to have right answers. We're supposed to be able to work them out and arrive at an agreed-upon solution. I have tried to show that this is not as easy as it looks, even in the realm of mathematical puzzles. Some of Paulos's problems do not have easy, definite solutions, no matter what he claims. Furthermore, in the example posed to Marilyn Vos Savant, most of her "numerate" audience kept insisting on the wrong solution to the problem. This world of mathematical certainty may not be so certain after all.

Furthermore, Paulos would have us believe that the abilities to be numerate are also linked to the abilities to be logical. I have tried to disprove this argument on at least two counts. First, I have noted that there is often no agreed upon "logic" to how even mathematics problems are solved. There are difficulties in defining solutions to some problems and no set answers to others. Second, many real problems are not so easily framed into logical or mathematical criteria. The accepted premises that underlie math and logic are not always so evident in everyday life.

The mystique of mathematics, however, is hard to overcome. As stated earlier in this chapter, there are strong beliefs that everyone should know a great deal of mathematics in order to function well in a democratic society. These beliefs touch many aspects of how we deal with mathematical thinking, the teaching of mathematics, and the ability to do mathematics. This book will examine many of these beliefs.

We have already challenged some of Paulos's beliefs about the supremacy of mathematics and have discussed the impracticality of many of his "simple" problems. To continue our investigation into the usefulness of mathematics, we must now return briefly to high school.

Notes

1. John Allen Paulos, *Innumeracy: Mathematical Illiteracy and Its Consequences* (New York: Hill and Wang, 1988).
2. Ina V. S. Mullis, John A. Dossey, Eugene H. Owen, and Gary

W. Phillips, *The STATE of Mathematics Achievement: Executive Summary* (Washington, D.C.: National Center for Education Statistics, 1991), p. 1.

3. John A. Dossey, Ina V. S. Mullis, Mary M. Lindquist, and Donald L. Chambers, *The Mathematics Report Card: Are We Measuring Up?* (Princeton, N.J.: The Educational Testing Service, 1988), p. 9.

4. John O'Neil, *Raising Our Sights: Improving U.S. Achievement in Mathematics and Science* (Alexandria, Va.: Association for Supervision and Curriculum Development, 1991), p. 5.

5. Galileo Galilei, *Dialogues Concerning the Two New Sciences*, trans. by Henry Crew and Alfonso de Salvio. In *Great Books of the Western World*, Robert Maynard Hutchins, ed. (Chicago: Encyclopedia Britannica, 1952), vol. 28, p. 190.

6. Paulos, *Innumeracy*, p. 7.

7. René Descartes, *Rules for the Direction of the Mind*, trans. by Elizabeth S. Haldane and G. R. T. Ross. In *Great Books of the Western World*, vol. 31, pp. 2–3.

8. Francis Bacon, *Advancement of Learning*. In *Great Books of the Western World*, vol. 30, p. 46.

9. Leo Tolstoy, *War and Peace*, trans. by Louise and Aylmer Maude. In *Great Books of the Western World*, vol. 51, pp. 47–48.

10. Paulos, *Innumeracy*, p. 11.

11. Ibid., p. 69.

12. Ibid., p. 51.

13. Ibid., p. 63.

14. Ibid., pp. 78–79.

15. Ibid., p. 79.

16. Ibid., p. 73.

17. Ibid., pp. 7, 27, 24, 37.

18. Lisa J. Evered, "Review of Innumeracy: Mathematical Illiteracy and Its Consequences," *The American Mathematical Monthly* 97 (1990): 89.

19. Ibid., p. 89.

20. Ibid., p. 90.

21. Ibid.

22. Ibid., p. 91.

23. Marilyn Vos Savant, "Ask Marilyn," *Parade Magazine*, September 9, 1990, p. 15.

24. Marilyn Vos Savant, "Ask Marilyn," *Parade Magazine*, December 2, 1990, p. 25.

25. Marilyn Vos Savant, "Ask Marilyn," *Parade Magazine*, February 17, 1991, p. 12.

26. Thomas Alva Edison, "Autobiographical." In *The Diary and Sundry Observations of Thomas Alva Edison*, ed. D. D. Runes (New York: Philosophical Library, 1948), p. 43.

27. Paulos, *Innumeracy*, p. 62.

28. Thomas Kuhn, *The Structure of Scientific Revolutions* (Chicago: University of Chicago Press, 1970).

29. Ibid., pp. 172–73.

30. Paulos, *Innumeracy*, p. 63.

31. Howard Gardner, *Frames of Mind: The Theory of Multiple Intelligences* (New York: Basic Books, 1983).

32. Ibid., p. 69.

33. Ibid., p. 135.

34. Ibid., p. 139.

35. Ibid., p. 167.

36. Ibid., pp. 167–68.

37. Ibid., p. 168.

2

Why Is Pythagoras Following Me?[1]

Could you be admitted to a college or university today? Remember the butterflies in your stomach on the morning you took a "college entrance" test? Remember your scores? More importantly, do you remember all the things you had to know in order to earn those scores? Before you continue reading this chapter, stop and try the sample test on the following page.[2]

If you read right on to this paragraph without attempting the test on the next page, please be advised that your opinions on the reform of education may be suspect. If you did take the test, please return your answers to me, and your score will be mailed to you in six weeks.

Just kidding. The answers to the test questions appear later in this chapter. While reading on in search of them, perhaps you will notice some of the serious criticisms I am making about mathematics education and education reform in general. If this chapter seems a little mysterious so far, perhaps you will be better prepared to understand the feelings that many students experience in dealing with mathematics. For, in a sense, school

Sample Test

DIRECTIONS

Solve each of the problems in this section using any available space for scratchwork. Then decide which is the *best* of the choices given and circle the letter that corresponds to your choice.

Note: Figures which accompany problems in this test are intended to provide information useful in solving the problems. They are drawn as accurately as possible EXCEPT when it is stated in a specific problem that its figure is not drawn to scale. All figures lie in a plane unless otherwise indicated. All numbers used are real numbers.

Please begin work. You have five minutes.

1. $(\sqrt{128})(\sqrt{32}) =$ $2\sqrt{32}$

   A. 64
   B. $32\sqrt{2}$
   C. $64\sqrt{2}$
   D. 128
   E. $64\sqrt{4}$

2. If $x^2 - 6x = -9$, then $x =$
   $(x-3)^2 = 0$

   A. -3
   B. 0
   C. 1
   D. 2
   E. 3

3.

B y x
A $4x$
$3x$ C
$4x$
D

ABCD is a quadrilateral (not drawn to scale).

$12x = 360$
$x = 30$

$y =$

   A. 30
   B. 60
   C. 90
   D. 150
   E. 180

4.

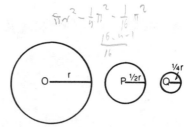

$\pi r^2 - \frac{1}{4}\pi^2 - \frac{1}{16}\pi^2$

$\frac{16 \cdot 4 - 1}{16}$

The area of circle O minus the area of circle P minus the area of circle Q equals

   A. $16\pi^2$
   B. $11\pi^2$
   C. $11/16\pi r^2$
   D. $12/16\pi r^2$
   E. $15/16\pi r^2$

5. Sam drove part of a trip at 70 miles per hour, and Bill drove the other part at 50 miles per hour. If Sam drove for 2 hours and Bill drove for 1 hour, what was their average speed (in miles per hour) for the entire trip?

140 $\dfrac{190}{3}$

   A. 50
   B. 60
   C. 63 1/3
   D. 66 2/3
   E. 70

   6

STOP

IF YOU FINISH BEFORE TIME IS CALLED, YOU MAY CHECK YOUR WORK ON THIS SECTION ONLY. DO NOT WORK ON ANY OTHER SECTION IN THE TEST.

DO NOT START READING THE REST OF THIS ARTICLE UNTIL THE FULL FIVE MINUTES HAVE ELAPSED.

Figure 1

mathematics is mysterious, and the current reform movements, which stress the need of every student for more mathematics, seem only to heighten the mystery.

Let me be clear from the outset. The mathematics requirements in our nation's high schools should not be increased, as so many proponents of education reform proclaim. In fact, they should be decreased. Why so many people support the former proposition and so few consider the much more reasonable and logical alternative of the latter is the theme of this chapter. Specifically, I will challenge the need for requiring between two and four years of abstract mathematics in our high schools. This requirement includes two years of algebra, one year of geometry, and one year of trigonometry and (perhaps) beginning calculus. I will focus on algebra and geometry because, whenever critics complain of declining scores on math tests, high school students generally end up taking more of these courses.

Why teach mathematics at all? How can we justify to our children the learning of such material? The rationale for mathematics instruction rests on three virtually unchallenged assumptions. The first, bequeathed by Greek philosophers and supported by modern science, is that mathematics is the foundation on which the universe is structured. Mathematics thus becomes philosophy and offers a methodology that will reveal the elegant harmony that seems to underlie the physical and, perhaps, even the social world. The second assumption, which draws its strength from a naive brand of psychology, assumes that studying the logical, deductive, and inductive methods common to mathematics will enable students to think more clearly and more rationally about other subjects (see chapter 1). This transfer-of-training argument hinges on the belief that understanding and practice in one area of knowledge will transfer to the learning of another. The third and more pragmatic premise is that mathematics is so frequently used in daily life that it should be taught to our children.

The evidence supporting each assumption is far from clear. In fact, in many cases, the evidence supports exactly the opposite conclusions about mathematics. In the case of the first assumption, the correlation between simple numbers and the physical world—the Greeks notwithstanding—has not been substantiated. The field of mathematics itself has splintered into numerous disciplines, each postulating a different view of the relationship between mathematics and reality. Second, the psychological evidence for transfer of training not only does not support this assumption but confirms its opposite: that training in one area may actually interfere with the learning of another. Finally, no study has supported the contention that the abstractions of algebra, geometry, and trigonometry, which so many students are required to learn, are practical in any general sense, except for a small number of occupations.

When all three assumptions—the philosophical, the psychological, and the pragmatic—fail and still students are forced to learn more abstract high school mathematics, I claim the following: the reform movements we are now encountering—especially calls for national standards and national tests—will not improve our children's mathematical abilities. Indeed, they may succeed only in creating a new generation that is either hostile to mathematics or anxious about dealing with it or both. How has this situation come about?

Math as Philosophy

To most adolescents in high school, *math* is algebra and geometry. In fact, we might argue that most adults, when asked an opinion about mathematics, will discourse on their experiences in one of these two subjects. Obviously, there are many different branches of mathematics. But during the impressionable teenage years, the axioms, postulates, and logical rigors (or hazards) of these two particular areas tend to leave an indelible impression about the nature of mathematics. Why algebra and geometry? Why even teach mathematics? To answer these questions we must discuss the centuries-old fascination with the mysteries of mathematics.

We must come to know Pythagoras (c. 582–c. 500 B.C.) quite well. Since the days of the ancient Greeks, mathematics has been seen not only as useful and playful but also as possessing a certain kind of power, a power that might provide a way to unlock the mysteries of the universe. Plato's academy in Athens was reported to have had a sign above its entrance proclaiming, "Let no one enter here who does not know geometry." Plato echoed and extended what, even in his day, was beginning to be a commonplace idea: that number is the matter and form of the universe. The Pythagoreans rejoiced in the fact that "all things are number" and built a mystical cult around the worship of numbers.

Why such excitement? How could numbers that could be counted on your fingers or simple figures drawn in the sand initiate a philosophical approach to nature that would last more than 2,000 years?

The explanation lies partly in the properties of numbers and figures themselves and partly in their remarkable correspondence, as it seemed to the Greeks, to things in the real world. Every student in high school has probably had an encounter with the Pythagorean theorem, which states that, given a right triangle, the square of the hypotenuse is equal to the sum of the squares of the opposite two sides ($A^2 + B^2 = C^2$). The Pythagoreans proved that this principle would hold for all right triangles, thus discovering a pattern that should transcend all times and all spaces. No wonder the Pythagoreans, when they found similar properties for squares, circles, and spheres, went on to proclaim that all nature is structured numerically.

These regularities probably would have remained only stimulating and interesting curiosities had it not been for the "discovery" that number could possibly describe the regularities of the physical universe. By applying numbers to the realm of astronomy, the Pythagoreans felt that the motion of planets and stars could be described mathematically. This connection between mathematics and the physical world inspired the first "standardized" school curriculum: the quadrivium, centered on four subjects linked by mathematics (arithmetic, geometry, music, and astronomy). Mathematical texts—particularly Euclid's *Elements*, the formalization of the propositions of geometry that is still at the heart of high school courses more than twenty centuries later—were revered and have been translated anew in each succeeding age. The mathematical method (and, by association, the philosophical and scientific methods), based on the deductive logic of geometry as codified by Euclid, cast a shadow over Western civilization that is perhaps rivaled only by that of Aristotle or the Bible.

Abstractions and regularities were seen to lie behind natural phenomena, and these nontransient facts became the focus of philosophy and science for the next several centuries. Plato set the tone in *The Republic*: "The knowledge at which geometry aims is knowledge of the eternal, and not of aught perishing and transient."[3] These eternal verities which had been designed by nature or the gods were the proper goal of all investigation. Twenty centuries later, Johannes Kepler would argue that "the chief aim of all investigations of the external world should be to discover the rational order and harmony which has been imposed on it by God and which He revealed to us in the language of mathematics."[4] The rationalist philosopher, Descartes, also echoed the belief in math as the supreme truth. In his *Meditations*, in arguing for innate eternal forms (much as Plato had done), he commented, "I counted as the most certain the truths which I conceived clearly as regards figures, numbers, and other matters which pertain to arithmetic and geometry, and, in general, to pure and abstract mathematics."[5] Finally, Galileo would argue for a nature that was designed mathematically:

> Philosophy (nature) is written in that great book which even lies before our eyes—I mean the universe—but we cannot understand it if we do not first learn the language and grasp the symbols in which it is written. The book is written in the mathematical language, and the symbols are triangles, circles, and other geometrical figures, without whose help it is impossible to comprehend a single word of it; without which one wanders in vain through a dark labyrinth.[6]

Math was to be the Rosetta stone for deciphering nature, the key that would reveal God's handiwork. And the key was fashioned according to the laws of algebra and geometry. For a time this seemed to be true: Kepler discovered three laws that explained planetary motion better than it had ever been done before; Galileo expressed mathematically the motion of devices like pendulums and clocks; and the supreme philosopher-scientist of all, Isaac Newton, uncovered what seemed to be three laws of the universe.* In Newton's *Mathematical Principles of Natural*

First Law: If a body is at rest or moving at a constant speed in a straight line, it will remain at rest or keep moving in a straight line at constant speed unless it is acted upon by a force.

Second Law: The time rate of change of velocity or acceleration is directly proportional to the mass of a body.

Third Law: The actions of two bodies upon one another are always equal and directly opposite.

Philosophy (1687), the paradoxes and past progress of all his predecessors were combined and reformulated into an ambitious system that seemed to account for everything. Mathematical designations could be made of the falling of a stone, the ocean tides, the motion of the moon, the precession of the equinoxes, and the movements of comets, planets, and stars. No wonder Pierre Simon Laplace remarked that Newton was fortunate because there was only one universe and Newton had discovered its laws!

This naive view began to change in the 1800s, and the changes that occurred not only rocked the foundation of mathematics but also made a mockery of claims that the structure of nature, the universe, society, or anything else is quite so simple—or, indeed, is even describable by mathematics at all. The simple nature of algebra and geometry yielded to a more complex mathematics, and doubts were raised about the efficiency of any mathematical descriptions of the world. Morris Kline commented on these changes in *Mathematics: The Loss of Certainty:*

> The current predicament of mathematics is that there is not one but many mathematics and that for numerous reasons each fails to satisfy the members of the opposing school. It is now apparent that the concept of a universally accepted, infallible body of reasoning—the majestic mathematics of 1800 and the pride of man—is a grand illusion.[7]

How did this happen? In general, the decline has come about on two fronts: first, the strictly logical foundations of arithmetic, algebra, and geometry have been challenged to such an extent that there are now several competing algebras and geometries; second, the dream of describing the world with one unified mathematics has been seriously questioned as different mathematics have come to describe different realities. In short, the belief in a single objective reality, describable by a single logical and truthful mathematics, which is based on algebra and geometry, has simply not been supported.

Certain highlights in the decline of mathematics should reveal some of the past and present problems. One of the first debacles came as nineteenth-century mathematicians challenged the logical structure of Euclid's geometry. What had been considered as logical and rigorously deducted axioms and theorems were shown to contain certain flaws in reasoning. In particular, Euclid's famous fifth postulate, that concerning two parallel lines, was shown to be open to different interpretations (i.e., parallel lines that supposedly would never cross might eventually meet in a curved time-space). Through the work of Karl Gauss, Nikolai Lobachevsky, James Bolyai, and Georg Riemann, a non-Euclidean geometry was constructed which posited almost a different view of reality: curved spaces, triangles whose angles add up to less than 180, and an infinite number of lines parallel to a certain line. In describing the world, Euclid became only one of the contenders instead of the unopposed champion.

Problems with simple numbers also led to the decline. Where the Pythagoreans believed that the world was structured as whole numbers or as ratios of whole numbers (such as 1, 2, 3), now different conceptions of numbers emerged for different purposes. For instance, irrational numbers, such as the square root of 2 (which had been known since Pythagoras) proved to be a problem for the logical foundation of arithmetic. What did it mean to have a concept that could not be explained in terms of simple numbers? Also the concept of negative numbers was highly problematic. Such numbers were useful in certain types of real-world thinking: for example, in life insurance tables where a negative number was seen as a debit. But how, in reality, could there be a negative quantity? Finally, imaginary numbers, like the square root of −1, useful in certain calculations, also seemed to boggle the logical mind. Are these concepts sensible or nonsensical? As William Freud declared in his *Principles of Algebra* (1796):

> (A number) submits to be taken away from a number greater than itself but to attempt to take it away from a number less than itself is ridiculous. Yet this is attempted by algebraists who talk of a number less than nothing; of multiplying a negative number into a negative number and thus producing a positive number; of a number being imaginary. Hence they talk of two roots to every equation of the second order (degree), and the learning is to try which will succeed in a given equation; they talk of solving an equation, which required two impossible roots to make it soluble; they can find out some impossible numbers, which multiplied together, produce unity. This is all jargon, at which common sense recoils; but, from its having been once adopted, like many other figments, it finds the most strenuous supporters among those who love to take things upon trust and hate the color of a serious thought.[8]

During the late nineteenth century and into the twentieth century, the challenges continued. The construction of matrix algebra showed that there were certain algebras in which the supposed immutable law of commutation did not hold, e.g., $A \times B \neq B \times A$. In the 1930s Kurt Gödel proved that no logical system complex enough to deal with numbers could ever be complete; that is, there may always exist statements that cannot be proved. Furthermore, the world, which was thought to be so rigorous, logical, and deterministic, was shown, at times, to be purely probabilistic. Probability theory came to be used more and more in areas ranging from particle physics to economics to psychology. And, finally, there was the computer. The logic and discrete mathematics of this tool proved to be different from simple algebra and geometry.

The foregoing historical digression is not meant to imply that the mathematics of algebra and geometry are inapplicable to anything in real life. Nor is it meant to suggest that any mathematical description is inappropriate. The efficiency of mathematics in describing certain natural phenomena and technological processes is unquestionable. I do mean to suggest, however, that the grand vision of a world structured according to geometry and algebra can no longer be accepted. As knowledge has progressed and as the world has changed, the study of mathematics has become more complex, and its ability to simply describe all phenomena has become more questionable.

But high school has not changed: the ancient Greek view of mathematics as algebra and geometry is still the core of the mathematics curriculum. And even when alternatives are taught, their philosophical implications are seldom discussed. More calculation and more theory are given, but rarely is the human struggle with mathematics presented. Mathematics is mysterious, but only in the sense that it reflects the continuing human attempt to solve the mystery of the universe. Studying the history of mathematics and the various modern attempts to model the world mathematically would teach students valuable lessons: that knowledge has limits, that mysteries remain in the universe, and that humankind is engaged in a never-ending attempt to unravel those mysteries. Isn't this what students should learn from school if we want to train them for future excellence?

Math and Psychology

The nature of the human mind has always been as mysterious as the fascination with mathematics. Perhaps this explains in part why the two have so often been linked. One of the most enduring notions that have been used to explain the mystery of the mind has been that it is composed of "faculties" or "powers" that, like the biceps, must be exercised to develop their strength. From the Greek philosophers to Descartes to the faculty psychologists and phrenologists of the nineteenth century to certain modern cognitive psychologists, the mind has been divided into elementary powers: reason, memory, language, logic, attention, perception, etc. To improve reason, a very important attribute of mind, one must exercise this faculty on the most difficult training equipment available: namely, mathematics.

For instance, the philosopher John Locke wrote:

> Would you have a man reason well, you must use him to it betimes, exercise his mind in observing the connection of ideas and following them in train. Nothing does this better than mathematics, which therefore I think should be taught all those who have the time and opportunity, not so much to make them mathematicians as to make them reasonable creatures.[9]

The psychological endorsement for the primacy of mathematics as a separate mental power has come from several sources. The early phrenologists specified a part of the brain that dealt with numerical intellectual powers. The construction of intelligence tests in America, by Lewis Terman and his associates at Stanford (the test now known commonly as the Stanford-Binet), had several number tasks as integral to the determination of an I.Q. score. L. L. Thurstone, in his influential work on psychometrics in the 1930s, postulated seven primary mental abilities which were thought to underlie all other abilities. Using a technique known as factor analysis, Thurstone identified a number skill (fundamental arithmetic operation) as one of his seven primary abilities. More recently, Howard Gardner, in his book *Frames of Mind,* has once again argued for a logical-mathematical ability as one of his seven "special" intelligences.

These philosophical speculations might have remained merely academic exercises had it not been for the profound practical implications of the belief in the primacy of mathematical thinking and of the assumption that ability or training in mathematics will transfer to other areas. For generations of college students, this belief has played an important role in determining their futures because a primary instrument used to judge admissions to college programs is the Scholastic Assessment Test (SAT), which includes mathematics as *one-half* of the total score. The only support for such a heavy emphasis on mathematics in this and other tests is the enduring belief that mathematical abilities are crucial for the success of all undergraduate and graduate students. Those who design and use the test (college admissions officers included) appear to believe that skill in mathematical reasoning should transfer to other areas that require logical thinking.

Such was the original belief of Carl Campbell Brigham. Working at Princeton from the mid-1920s to the late 1930s, Brigham developed several versions of the SAT. The one central purpose of his work was to assess a student's *general ability,* independent of specific cultural influences or peculiar high school training programs, in order to provide a test that would rank students from throughout the country. The test score came to consist of two major parts: verbal and mathematical. The SAT displayed good reliability (student scores did not seem to fluctuate much on retests) and fair validity (the test could reasonably predict college grades). From the verbal and mathematical abilities students demonstrated on such a standardized test, colleges and universities would be able to select students of outstanding ability and future promise. In at least this one context, the dictum of

Plato that no one would succeed without geometry has proved to be true. But is this belief in the transfer of training justified?

The idea of a strong degree of transfer of training, based on mental faculties, runs deep in the mental psychologies and philosophies of the nineteenth century. However, no empirical evidence exists to support such a claim of transfer of training, especially the broad proposition that the learning of mathematics will facilitate, say, the learning of the logic of physics or the logic of economics.

One of the earliest pieces of negative evidence comes from the celebrated psychologist William James. Using himself as subject, James tried to determine whether or not memorizing poetry would enhance his ability to memorize in general. First, he learned 158 lines of Victor Hugo's *Satyr* and kept a careful record of the time required. Next, he spent more than a month committing Milton's *Paradise Lost* to memory. Finally, he memorized another 158 lines of *Satyr*, which required more time than did the first 158 lines. James concluded that all his practice on *Paradise Lost* had not improved his "memory faculty"; in fact, he wasn't any better at memorizing poetry, the subject he had practiced on, much less at memorizing other different material.

In an important paper published in 1901, Edward Thorndike and Robert Woodworth reported finding no evidence that improving one mental function would improve another. In their experimental tests of the effects on one another of such concepts as discrimination, attention, memory, and other mental functions, they found no evidence for a transfer of training. They concluded:

> Improvement in any single mental function rarely brings about equal improvement in any other function, no matter how similar, for the working of every mental function group is indicated by the nature of the data in each particular case.[10]

In a conclusion similar to James's disparaging observations about the effect of memorizing *Paradise Lost* on memory in general, Thorndike and Woodworth wrote that, "improvement in any single mental function need not improve the ability in functions commonly called by the same name. It may injure it."[11]

In the 1920s Thorndike moved more directly into research on the learning of mathematics, especially the learning of arithmetic and algebra. In his 1923 volume on the *Psychology of Algebra,* Thorndike and his fellow researchers at Teachers College, Columbia University, criticized the techniques behind the teaching of the "old math":

> The faith in indiscriminate reasoning and drill was one aspect of the faith in general mental discipline, the value of mathematical thought for thought's sake and computation for computation's sake being itself so great that what you thought about and what you computed with were relatively unimportant.[12]

Drawing support from the belief in mental faculties, the "old math" had emphasized the value of mathematical reasoning in its own right, as a subject whose rudiments could easily transfer to the learning of other disciplines. Thus, algebra textbooks could deal with "bogus" and "fantastic problems" because the content wasn't nearly as important as the general principle behind the content. But does performance on these types of problems, or math in general, transfer to other domains?

It is well-known that scores on intelligence tests, including scores on the mathematics subtests of such standardized measures as the SAT, correlate well with academic grades. There is very little evidence, however, that they correlate with much else—particularly with those things that might predict success in later life. In a review of the relevant research, Michael Wallach notes that tests seldom predict real-life *accomplishments*. In a study of five hundred undergraduates, Wallach found that original accomplishments outside the classroom—in literature, science, art, music, dramatics, political leadership, and social science— were unrelated to SAT scores, although high SAT scores were related to high grades. Citing a study by Lindsey Harmon, which interrelated the professional contributions of physicists and biologists with data on the academic proficiency of college students, Wallach wrote, "How good a professional scientist the person became could not be predicted from any of this academic proficiency information." Should we then have stronger test standards? Wallach *concludes:*

> The irony is compounded when educational commentators . . . urge even heavier use of academic aptitude test scores instead of grades in selection on the ground that the former are less subject than the latter to irrelevant sources of bias such as teachers giving higher grades to students who are more polite. For it is their correlation with grades that provides the aptitude tests with their basic justification in the first place.[13]

Yet the belief in the power of mathematics as an aid to logical thinking continues. As a popularly held notion, we can perhaps countenance such a stand, for popular notions often reflect current fancy. As a professional opinion, however, we must be less tolerant of ill-founded notions. Educators and scholars should vigorously examine the validity of any idea that has such profound effects on the lives of children. But even among professionals, there is a wide difference of opinion concerning the power of mathematics.

In the early 1970s, Ohmer Milton of the Learning Research Center at the University of Tennessee surveyed assistant, associate, and full professors and members of the professional staffs on several college campuses. He asked the respondents to agree, disagree, or say that they were undecided about the truth of the statement: "The study of mathematics is especially important in helping students to think logically." Sixty-two percent of all professors in natural, biological, and physical sciences agreed with the statement; only 18 percent disagreed.

The trend among faculty members in other disciplines and among members of professional staffs was exactly the opposite. Only 26 percent of professors in the humanities agreed; only 32 percent of those in the social sciences agreed; only 40 percent of members of professional staffs agreed. The only truly strong support for the generalizability of training in math comes from the "hard" sciences—those professionals who have longer histories of using mathematics as models for research. For other university people, the belief that mathematics is a tool that disciplines the mind is not widespread. As Milton concluded, "The fact that so many of the 'hard' scientists believe in such a fundamental issue in the absence of the kinds of data they themselves would accept may be presumptive evidence against automatic subject matter transfer."[14]

The notion that the study of mathematics will help in the study of other subjects is simply erroneous. There is no experimental evidence to support the broad idea of transfer of training. Given that math is not the key to the universe or the key to learning other material, if math is still to be universally taught, then only one argument remains: math must be very practical and useful. As Thorndike hoped,

> emphasizing ability to use algebra in solving problems which life will offer, it seems desirable to consider the lives of boys and girls and men and women as students, citizens, fathers and mothers, lawyers, doctors, business men, or nurses and

select problems which they may usefully solve and which are properly solved by algebraic methods.[15]

Math and Practicality

Is the math that is now being taught in the nation's high schools useful in the lives of the ordinary citizens of this nation, as Thorndike hoped it would be? If we can demonstrate such practicality for the mathematical topics that are normally covered in high school, then an argument for retaining these courses can be made. If not, we must consider why math is taught at all, and what effect the teaching of something so impractical has on the students who are forced to take it.

In the following discussion, remember that I am talking about the high school mathematics quadrivium—algebra I and II, geometry, and either trigonometry or precalculus. Let me not deal in abstractions; those who continue to suggest that "math skills" are declining and that "math standards" must be raised should bear in mind how these suggestions are translated in the textbooks used in high school classes.

Even in the 1920s, Thorndike complained of the "fantastic" and "bogus" problems that were used in high school algebra textbooks. These problems were usually organized by the algebraic technique involved and not according to their use in real life. If we were to scan the textbooks used in these courses today, I think that we would find that the situation hasn't changed much from the turn of the century.

For example, let's investigate the teaching of algebra. In a detailed analysis of topics in algebra, Thorndike felt that only a few could be called basic principles: the idea of symbolism; the ability to read formulas; the ability to evaluate and solve formulas, first with one unknown and then with two or more; and, finally, the ability to read graphs. How many of us who have had two years of algebra could tell someone else so succinctly what the essentials of algebra are?

It should not be surprising if we failed. For from these simple notions, students are drilled in countless variations on the basic themes. The ability to read formulas, for example, might come in handy if we are asked to evaluate interest rates, to convert degrees Fahrenheit to degrees Celsius, to convert miles to kilometers, or, perhaps, even to solve a physics problem. However, endless pages of drill are devoted *not* to such practical matters but to the evaluation of such equations as:

$$Divide\ y^3 - 6y^2 + 14 - 12$$
$$by\ y - 3$$

$$Or\ multiply\ \frac{2a^2 - a - 28}{3a^2 - 9 - 2}$$

$$by\ \frac{3a^2 + 11a + 6}{4a^2 + 16a \times 7}$$

Homework often consists of solving endless variations on a simple principle. Those who feel that I am exaggerating should talk to their children.

Most textbooks do attempt to provide some practical applications of these concepts through word problems. All of us who teach any kind of postsecondary mathematics know the shudders that run through our classes when "word problems" are mentioned. Why? A brief examination of any algebra textbook will reveal that these attempts at practicality often become exercises in absurdity. Consider these examples:

Sam can do a job in five hours working by himself. Bill, however, can do it in three hours. How long will it take them if they work together?

*   *   *

Miguel is eight years older than his brother, Juan. Five years from now, he will be twice as old as Juan. How old are the boys now?

*   *   *

Jonathan can bicycle from his home to school in 12 minutes while his older sister Suzanne can make the same trip in eight minutes. If Jonathan starts from home to school at the same time that Suzanne starts from school to home, how long will it be before they meet?

Although all three examples have "algebraic solutions," how sensible are the assumptions that we must make to find the correct answer? When two people work together on a task, simple psychology indicates that the time to complete it depends on many factors. Similarly, when two people start out on a trip, a time when they will meet is contingent on several things. For anyone who has lived a normal life, these last two statements are only common sense. Finally, concerning the age problem, how many people would not know the ages of their own brothers?

The fact that most word problems seem to involve purely mathematical techniques or tricks would be fine if they didn't masquerade as real-life, applied concepts of algebra. The silliness of these exercises fosters a sense of absurdity in students (for more see chapter 3). The techniques they are learning are only nonsensically applied, so why should they bother to learn the material? As Thorndike suggested, solving problems in school ought to be for the sake of solving problems in life. Thus problems should be organized to match techniques with real-life applications.

Now we might ask, just how much of the high school mathematics quadrivium can be used in everyday situations? Once again Thorndike is a good guide. In his 1923 book he examined a sample of articles from many sources—encyclopedias, almanacs, and so on—to see how much algebra was actually incorporated into real-world settings. His conclusions were mostly negative: only a fraction of the high school algebra that was then taught was actually used. Is this surprising? From our own experience, how often do we use the math that we learned in high school?

I think these techniques are used only under certain limited conditions in certain jobs—mostly engineering and other technically oriented occupations. Must all students endure four years of math for the sake of producing a few technicians?

In one critical respect, the answer to this question is a definite yes, for nearly all high school students who plan to attend college are forced to deal with math in a way that affects their futures. I am referring once again to college admissions tests. In most cases, the math required on these tests is high school algebra and geometry.

Could you be admitted to college today? If you didn't take the test at the beginning of this chapter, now is a good time to do so. If you did, the answers are as follows: 1. A; 2. E; 3. D; 4. C; 5. C. To determine your admissions status, use the following college admissions key (approximate, of course): one correct (no admission); two correct (admission with remedial math work); three correct (general admission); four correct (admission with a good chance at a scholarship); five correct (admission with a great chance at a scholarship or fellowship).

Given this approach, students see math as nonsensical material that they are coerced into learning, not as a freely chosen, inherently interesting subject. And coercion has its own deterrent effects, as Einstein observed:

> One had to cram all this stuff into one's mind for the examinations, whether one liked it or not. This coercion had such a deterring effect on me that, after I passed the final examination, I found the consideration of any scientific problems distasteful to me for an entire year.[16]

Math Anxiety

So far in this discussion, I have tried to shatter the assumptions that support educators' reliance on the teaching of mathematics: that mathematical models reflect in a simple way the structure of reality; that training in mathematics will facilitate training in other subjects; and that mathematics, as defined by high school courses, is useful in the workaday world. If all of these are unfounded assumptions, what justification remains for requiring—and perhaps even increasing the requirements for—these subjects? More important, what effects might we expect these increased requirements to have on high school students, forced to deal even more intimately with mathematics?

Reasonable criteria do exist for increasing the amount of mathematics available to students in high schools. We read constantly of our nation's need for technical expertise, for more scientists, engineers, and computer scientists to help keep the country strong in research and development and a continued world leader in technology. To this end, sophisticated and abstract courses in science and mathematics must be *offered* in both high school and college.

However, since only a small minority of high school students actually enter these professions or even wish to do so, must all high school students be *forced* to endure a highly specialized mathematics curriculum in order to select the few who are needed? Must all students, whatever their future jobs, be forced to deal with two to four years of such highly abstract technical subjects as algebra and geometry, which will be useful only in a relatively small number of professions?

Such is indeed what happens and will continue to happen at an accelerated pace if the current proposals for revising high school curricula are adopted. An even greater number of high school students will have to endure even more algebra and geometry in order to graduate. Those who are more inclined toward the liberal arts and who see no future at all in mathematics must either learn the subject or accept the reduced chances of admission to and scholarships from major universities that are the direct result of lower high school grade-point averages and lower scores on college admission tests.

For those who encounter math, endure it, but don't like it, we might expect "math anxiety" to develop, a fear that might keep them away from any math-oriented disciplines. Recent research on math anxiety has shown the phenomenon to be real and to have an impact on people's perceptions of themselves and of their career choices.

In *Overcoming Math Anxiety*, Sheila Tobias describes the sexual biases that are built into mathematics curricula (also see chapter 4).[17] Males are more highly reinforced for their mathematical abilities; the problems in textbooks are slanted toward a male perspective (males get the rewarding mathematical tasks: Johnny does math on the job, Susie in the kitchen); and even the behavior of teachers shows evidence of bias toward males in the sense of believing that males are more competent at math than females. Thus increasing the number of required mathematics courses not only favors future technicians and scientists, but also probably favors males who have traditionally chosen to pursue mathematically oriented careers in greater numbers than females. Obviously, this trend might change as more females enter these professions. But how many never get the opportunity to do so because of some early discouragement concerning their mathematical ability?

Certainly, not everyone will come to hate math. However, repeated exposure to math in high school, reinforced by the impracticality of the material, can only plant questions in the minds of even the most capable students: For what purpose? Is there a reason for all this, other than the mental gymnastics involved? If no answers to these questions emerge from students' later lives, then high school courses facilitate a "cult" of mathematics: that is, the belief grows that the material must have some value, somewhere, sometime. The student doesn't know where or when, but surely somebody must know. An air of mystery then surrounds mathematics, a numerical magic that keeps us from questioning the users of numbers, for we feel that what they do must be right, for we believe that mathematics is always right!

In forcing students to take such courses we are creating a generation of students more hostile to and more anxious about mathematics than preceding generations. And for the vast majority of them, the mystery and magic of numbers deepens.

What to Do?

I have tried to show that the rationale for the way we present mathematics in high school is implausible. We cannot continue to force our children to take ever-heavier loads in such subjects as algebra, geometry, and trigonometry, courses that even adults cannot justify in terms of practical significance. What, then, should be the place of mathematics in high school and beyond?

1. The history and philosophy of mathematics should be taught to all students. To eradicate the mystery that needlessly attaches itself to numbers, we must acknowledge their human attributes; that is, we must explain how and why people have used mathematics to understand their world. Perhaps we should call this a "math appreciation" course. In such a course, we could try to justify the study of math, rather than simply force everyone to study years of algebra and geometry.

2. The only math requirement in high school should be one year of basic math. This course would include practice in simple arithmetic skills, as well as training in the use of calculators and in such simple applications of mathematics as interest rates (as used by banks, savings and loan associations, department stores, etc.); mortgages and loans; and tax calculations. The use of real-life equations can be introduced and students can be taught the essential elements needed to solve them.

Furthermore, I think it is crucial that a section on "government math" be included. That is, students should receive an introduction to the ways in which large organizations (industry could also be included) arrive at various estimates, from gas mileage to defense spending. One of the major goals of schooling in a democratic society, as Jefferson foresaw, was the production of educated citizens. I can't think of anything more practical than lessons in how real-life mathematics works. (For an expanded discussion of this point of view, see chapter 7.) Students need the opportunity to learn real-life problem solving that involves the use of mathematics—not simply rote drill.

3. Algebra I, algebra II, geometry, and trigonometry/precalculus should become electives. No student should be required

to take these courses. Those who are inclined toward these subjects should certainly have the opportunity to study them in depth. Others who couldn't care less would better be served by more time in literature or history courses.

Some might claim that, by eliminating the requirement that all students take these math courses, we will risk overlooking a host of talented young people who might become researchers in technology and science. I claim just the opposite: more of these professionals would be produced, as long as the demand for them exists. High school students who show an early interest in mathematics and science will pursue these subjects as electives, and we needn't be concerned with these groups of students. From those who develop these interests later, say, in the first or second year of college, we might indeed recruit more engineers and technically oriented professionals. However, bad experiences in high school math can steer people away from mathematically oriented professions and prevent these late bloomers from following up on their interests. Eliminate the bad experiences, and we might well have more people choosing technical professions at a more mature age. Most universities could teach the prerequisite math necessary for beginning engineering students in a one- or two-semester sequence. If students in these courses want to be there, the material could probably be taught even faster.

When I teach statistics and research design to undergraduate students in psychology, I have often wished that my students had no high school mathematics. What they remember of the content of these courses is usually minimal. But they bring with them a crippling desire not to take any "math-related" course again. Pythagoras haunts them and affects how I must now teach them.

4. The math sections of the major standardized tests, such as the SAT, should be eliminated. Math is not that important to many professions, and to require all students to master it is

ludicrous. By doing so, we only deny admission to some otherwise capable students and make people in the tutoring business rich. Additionally, a national test in mathematics (see chapter 7) would also be disruptive.

The central issue is one of choice and freedom. We automatically assume that the cure for "declining standards" is more of everything: more hours of schooling, more homework, more math courses, and so on. In solving educational problems, we seldom stop to think about learning. We seldom ask how we are going to instill in our students the desire to learn and to create instead of just the ability to make the grade.

High schools are currently designed to foster confinement, not to encourage freedom. They consider constraint, not freedom, to be the essence of American education. In this way, our high schools hope to produce a new generation of talented individuals to invent new technologies, to discover new ideas, and to create new products to keep the country strong. But coercion does not encourage curiosity, as Einstein noted:

> It is, in fact, nothing short of a miracle that the modern methods in instruction have not yet entirely strangled the holy curiosity of inquiry; for this delicate little plant, aside from stimulation, stands mostly in need of freedom; without this it goes to wreck and ruin without fail. It is a very grave mistake to think that the enjoyment of seeing and searching can be promoted by means of coercion and a sense of duty.[18]

Freedom of choice is what has always made this country great. Isn't it time we put it back into our classrooms and gave it back to our children?

Notes

1. A slightly shorter version of the material in this chapter appeared in *Phi Delta Kappan* (February 1989): 446–54.

2. The examples in this test are taken from Michael K. Smith, Michael C. Hawthorne, and Karen Josvanger, *The SAT Video Review Study Guide* (Knoxville, Tenn.: CAM3 Associates, 1988), pp. 50, 57, 63–64, 67.

3. Quoted in Morris Kline, *Mathematics: The Loss of Certainty* (Oxford: Oxford University Press, 1980), p. 16.

4. Ibid., p. 31.

5. Ibid., p. 42.

6. Ibid., p. 46.

7. Ibid., p. 6.

8. Ibid., p. 154.

9. Thomas Fowler, ed., *Locke's Conduct of Understanding* (New York: Macmillan, 1980), p. 20.

10. Quoted in William S. Sahakian, *Learning: Systems, Models, and Theories* (Chicago: Rand-McNally, 1976), p. 100.

11. Ibid.

12. Edward L. Thorndike, *The Psychology of Algebra* (New York: Macmillan, 1923), p. 96.

13. Michael A. Wallach, "Tests Tell Us Little About Talent," *American Scientist* 64 (1976): 63.

14. Ohmer Milton, "Learning Transfer: A Diogenian Task," *Teaching-Learning Issues* 15 (1971): 8.

15. Thorndike, *The Psychology of Algebra*, p. 109.

16. Dean Keith Simonton, *Genius, Creativity, and Leadership* (Cambridge, Mass.: Harvard University Press, 1984), p. 63.

17. Sheila Tobias, *Overcoming Math Anxiety* (Boston: Houghton Mifflin, 1978).

18. Simonton, *Genius, Creativity, and Leadership*, p. 63.

3

The Word Problem

Two trains leave at the same time but from different cities. One train is traveling west at 120 mph while the other is traveling east at 100 mph. Assuming the trains are traveling on opposite but parallel tracks, and the two cities from which they leave are 440 miles apart, how long will it take before they meet?

It seems that everytime I mention the train problem, whether it be in classes, professional meetings, or personal conversations, nervous laughter ensues. Practically everyone remembers the difficulties of doing word problems in high school math courses. Why are these problems so difficult and why do they evoke such bad memories? Do people have latent phobias about trains, perhaps caused by a primal archetype that even Jung overlooked?

In fact, the mere mention of "word problems" can produce some heated responses from those who have been through math classes. I once asked a group of undergraduates to respond to the term "word problem" with whatever came to mind. Some typical comments:

> Don't like them because they make problems that are very simple very hard to understand. Torture—long unnecessary word problems which make no sense.

> Tricks to fool the student.

> Yuk! I never did care how fast two trains heading in opposite directions were going or which one would get wherever first.

> Confusion—sitting at a desk trying to figure out what they are asking—have nothing to do with anything.

> Panic! Who cares? If a train leaves from NYC (100mph) at 8:30 for Florida and a plane leaves at 2:30 Atlanta for NYC (300mph) when will the two be at the same point?

> How do you spell a psychotic shriek?

> I think of stupid fictional Johnny's riding across the town (approx. 8 miles) to get 8 apples for his mother and determining the cost of each apple.

> Annoying, time consuming—When the hell are the trains from Chicago and LA going to hit each other?

> Never seemed to serve any purpose, they were just assumed to be necessary. Two trains going in opposite directions. . . . Word problems are senseless pieces of junk derived for the sole purpose of getting revenge on high school students. They were formulated by a psychologist who wanted to test the amount of stress the high school mind could take before a nervous breakdown.

Not all responses are so negative. Some students see word problems as challenging and stimulating brain teasers. But the fact remains that a good percentage of students respond negatively to this type of mathematics. This leads to two questions: Why the negative response? Why are word problems included at all?

What is the purpose of the word problem? Why is it included in math textbooks and what is it supposed to be testing? If you take a look at any high school text in algebra, you'll probably notice that word problems are not as common as other exercises. Most problem sets feature countless variations on some simple principle, as evidenced by the following examples from a typical page of homework.[1]

$$\frac{r^2 + 4r + 3}{r^2 - 8r + 7} \times \frac{r^2 - 2r - 35}{r^2 - 7r - 8} \times \left(\frac{r^2 + 8r + 15}{r^2 - 9r + 8}\right)^{-1}$$

$$(z^2 + z - 2) \times \frac{z^2 + z - 20}{3z^2 - 2z - 1} \Big/ \frac{2z^2 - 5z - 12}{2z^2 + 3z}$$

$$\left(\frac{a^2 \quad ab - 2b^2}{3a^2 - 7ab + 2b^2}\right) \times (a^2 + 4ab) \times \left(\frac{a^2 + ab}{6a^2 + 7ab - 3b^2}\right)^{-1}$$

$$\frac{15 - 13y + 2y^2}{4y^2 - 9} \times \frac{2y + 1}{1 - 2y} \Big/ \frac{5 - y}{2y - 1}$$

$$\frac{30 - 11c + c^2}{9c - 6c^2 + c^3} \times \frac{c^2 - 3c}{25 - c^2} \Big/ \frac{c^2 - 9}{c^2 + 2c - 15}$$

Most texts view mathematics as symbol manipulation; learn, for instance, how $a^2 - b^2$ is factored, and practice this principle in endless drills which vary certain aspects of a particular technique.

When educational reformers talk about children doing more

homework each day, these types of exercises take on even more importance. I'll never forget a conversation with a bright math honors student from a good high school in Tennessee. The young man had made a perfect score of 800 on the mathematics portion of the Scholastic Assessment Test. When I asked how he did in high school math courses, he said his grades were good but very often he was bored by the assignments. I presented him with the exercises above and asked him to tell me how he would work on this assignment. He said that he would scan the page and look for what seemed to be the hardest problem and try to work it. If he failed, he would back up a bit until he understood the principle discussed. If he succeeded, however, he would stop, figuring he knew what was going on. Unfortunately, his teachers and parents didn't like this strategy and wanted him to *do his homework* and he had to work out pages of simpler problems.

The previous discussion is important in setting the context for word problems. Most math textbooks emphasize rote drill and abstract manipulation of mathematical techniques. After these exercises, word problems are presented. In a literal sense, we can see where they obtained their name—they are the only problems present that have *words* in them and not just symbols, such as xs and ys.

Word problems are supposedly real life applications of the mathematical principles under discussion. Since they are intended to expand upon the techniques presented, they naturally come at the end of a section. Learn the technique in the abstract and then practice it in more concrete applications. The intent, then, is laudatory: to show students how the mathematical principles they are learning are related to mathematical problems they will encounter in real life.

How well do word problems in typical texts accomplish this purpose? My examination of numerous high school texts leads me to the following conclusions:

1. Word problems reinforce the idea that abstract symbol manipulation is the essence of mathematical thinking, since these types of problems view words as being merely transparent facades. In other words, look past the words to the real math involved.

2. Word problems are *not real*. They are often gross simplifications of real life situations masquerading as genuine problems. As such, word problems seem hastily constructed and poorly thought out.

3. Word problems create an attitude that to do well in math is to be able to think quickly and abstractly, to ignore complexities, and to search for the right answer. They imply that the real world is structured simply, and unequivocally, in mathematical terms.

To begin illustrating these points, let's examine some typical word problems and how students attempt to solve them. In an influential article on word problems, Nobel laureate Herbert Simon and his associates Dan Hinsley and John Hayes presented "data on the comprehension of those popular twentieth-century fables called algebra word problems."[2] To see if these word problems could be grouped into distinctive "types," they asked students to sort a number of problems taken from algebra textbooks into piles, with each pile representing similar problems. They found that 84 percent of all high school algebra word problems could be classified into eighteen clusters. Examples from ten of the clusters are presented in table 3.[3]

Table 3

1. TRIANGLE

Jerry walks 1 block east along a vacant lot and then 2 blocks north to a friend's house. Phil starts at the same point and walks diagonally through the vacant lot coming out at the same point as Jerry. If Jerry walked 217 feet east and 400 feet north, how far did Phil walk?

2. DISTANCE
   RATE
   TIME

In a sports car race, a Panther starts the course at 9:00 a.m. and averages 75 miles per hour. A Mallotti starts four minutes later and averages 85 miles per hour. If a lap is 15 miles, on which lap will the Panther be overtaken?

3. AVERAGES

Flying east between two cities, a plane's speed is 380 miles per hour. On the return trip, it flies 420 miles per hour. Find the average speed for the round trip.

4. SCALE

Two temperature scales are established, one, the R Conversion scale, where water under fixed conditions freezes at 15 and boils at 405, and the other, the S scale, where water freezes at 5 and boils at 70. If the R and S scales are linearly related, find an expression for any temperature R in terms of a temperature S.

5. RATIO

If canned tomatoes come in two sizes, with the radius of one being 2/3 the radius of the other, find the ratios of the capacities of the two cans.

6. INTEREST

A certain savings bank pays 3% interest compounded semiannually. How much will $2,500 amount to if left on deposit for 20 years?

7. MAX-MIN

A real-estate operator estimates that the monthly profit p in dollars from a building s stories high is given by $p = -2s^2 + 88s$. What height building would he consider most profitable?

8. NUMBER

The units digit is 1 more than 3 times the tens digit. The number represented when the digits are interchanged is 8 times the sum of the digits.

9. WORK

Mr. Russo takes 3 minutes less than Mr. Lloyd to pack a case when each works alone. One day, after Mr. Russo spent 6 minutes in packing a case, the boss called him away, and Mr. Lloyd finished packing in 4 more minutes. How many minutes would it take Mr. Russo alone to pack a case?

10. NAVIGATION

A pilot leaves an aircraft carrier and flies south at 360 miles per hour, while the carrier proceeds N30W at 30 miles per hour. If the pilot has enough fuel to fly 4 hours, how far south can he fly before returning to his ship?

Later on I'll discuss why I think these problems are "not real." For now, let's continue with Simon and his colleagues' investigation of how students solve these typical word problems. The researchers next tested how quickly students could recognize a new problem as being one of these types. To do this, problems were read by the researchers in bits and students asked to categorize the problem and state what question was going to be asked. They found that over half of the students could correctly categorize the problem after hearing less than one-fifth of the text! In other words, the mention of "Two trains . . ." almost immediately evokes a recognition that this is some type of distance-rate-time problem for those students who have learned this particular algorithm.

Simon and his associates did find that students use standard information to solve a word problem, if they can recognize it as one of the types. If the problem is nonstandard, i.e., does not readily conform to one of the familiar clusters, then they adopt a line-by-line solving strategy. But students will attempt first to categorize if at all possible. This means that if a word problem can be recognized as a distinctive type, and a person has learned the algebraic equations that correspond to that type, then that persons stands a much better chance of solving it.

The solution processes involve very little creativity. Instead, the ultimate intent is: Can I recognize this problem as something that I've seen before so I can solve it? Simon and his associates found this to be true even if the problems were nonsense. For instance, to which category does the following problem belong and how would you solve it?

Chort and Frey are stimpling 150 fands. Chort stimples at the rate of four fands per yump and Frey at the rate of six fands per yump. Assuming that Chort and Frey stimple the same number of yumps, how many fands will Chort have stimpled when they finish?[4]

This was easily recognized as a variation on the work-type problem and could be solved by ratios. If the words are meaningless and yet the problem can be solved, what does this say about learning mathematics? This confirms point 1 above: word problems are afterthoughts to the mathematical idea that symbols and symbol manipulation are the most important things one can learn from math. It also reinforces the notion that the "real world" is irrelevant: as long as something can be recognized as a type of mathematics, it can be solved.

The second point argued that word problems are gross simplifications of real-life situations. Take the train problem, for instance. In real life, if two trains left separate stations at the same time, how long would it take for them to meet? I grew up around an uncle who worked for the L & N Railroad. He was stationed out of Knoxville and alternately did jobs as a brakeman, lineman, and conductor. When I was a child, he would take me on a passenger train from Knoxville to Cincinnati and back, in the days before the L & N discontinued passenger service. At the start of each trip, in my youthful impatience, I would ask him how long it would take to arrive. He always told me "trains move at their own speed"; after several trips, and several different arrival times, I began to understand the human comparison.

How realistic are some of the other types of problems in Simon's list? As a psychologist, I've always been interested in the "work" problems: questions such as, "If Sam can do a job in 5 hours and Bill can do it in 3 hours, how long will it take them working together?" Now, this can be calculated with an algebraic algorithm to determine an answer, precise to the second decimal place. But how useful is this? How long does it take two people together to finish a job? To me, this is a wonderful psychological problem: it involves factors of how well the people know each other, how important the job is, what time of day it is, what food and/or beverages are available, is the boss around, is it Monday or Friday, are they in a good mood, and a thousand other variables. Now, we can estimate the combined time to completion but even this guess involves making a number of assumptions about group cooperation and working conditions. Ask any employer! To imply that this question has a precise answer is misleading to those trying to solve it.

Or consider another type, which did not make Simon's list but which I have encountered quite often: the age problem. Sample question:

Sally is five years older than her brother Bill. Four years from now, she will be twice as old as Bill will be then. How old is Sally now?

First of all, who would ask such a question! Who would want to know this? If Bill and Sally can't figure it out, then this is some dumb family.

The impracticality of mathematics problems can be affirmed in a story told by the late Richard Feynman, Nobel laureate in physics. He was asked to serve on the mathematics textbook panel for the State of California, a group of educators, parents, and some professionals who are asked to help select textbooks for use in California public schools. To do his job well, he decided he needed to read all of the textbooks under consideration, an idea which he found was not adhered to by many of the other committee members. He ordered the books and a few days later three hundred pounds of them showed up!

As he was reading, he said he kept exploding like a volcano because,

> the books were so lousy. They were false. They were hurried. They would try to be rigorous, but they would use examples . . . which were *almost* OK, but in which there were always some subtleties. The definitions weren't accurate. Everything was a little bit ambiguous—they weren't *smart* enough to understand what was meant by "rigor." They were faking it. They were teaching something they didn't understand, and which was, in fact, *useless*, at that time, for the child.[5]

He cites an example of a word problem, trying to apply math to science. It read in part:

> *Red stars have a temperature of four thousand degrees, yellow stars have a temperature of five thousand degrees, green stars have a temperature of seven thousand degrees, blue stars have a temperature of ten thousand degrees, and violet stars have a temperature of . . . (some big number).*

As Feynman remarks, "There are no green or violet stars, but the figures for the others are roughly correct. It's *vaguely* right— but already, trouble! That's the way everything was: Everything was written by somebody who didn't know what the hell he was talking about, so it was a little bit wrong, always!"[6]

It got even worse when the question based on this information was posed:

> *John and his father go out to look at the stars. John sees two blue stars and a red star. His father sees a green*

*star, a violet star, and two yellow stars. What is the total
temperature of the stars seen by John and his father?*

Feynman reports that he would explode in horror: "Perpetual
absurdity! There's no purpose whatsoever in adding the temperature of two stars. Nobody *ever* does that except, maybe, to then
take the *average* temperature of the stars, but not to find the
total temperature of all the stars!"[7]

Feynman comments also on the pressures endured by members of the textbook panels. They were offered gratuities indirectly
by textbook companies, since the decisions of these committees
often meant millions of dollars to publishers. With so much to
read, many members would glance cursorily at the books under
consideration. This is not to say this happens everywhere,
everytime, but we can see the temptations: with so much money
riding on textbook adoptions, do publishing companies have the
luxury to produce high-quality, extremely well-written and well-
thought-out texts?

The third complaint about the present way that word problems are structured is as follows: Word problems imply that to
succeed in math we have to think quickly and abstractly, ignore
complexities and go for the right answer. This has two implications: it hints that to do well in math is also to think well
in general, and that most of the real world could be structured
simply and unequivocally in terms of mathematics.

Nowhere can this be seen more clearly than in the mathematics test that high school students take each year, the Scholastic
Assessment Test, or SAT.[8] SAT scores determine admission or
nonadmission to colleges and universities for thousands of high
school students each year. A poor score on this exam may over-
shadow good GPAs, good letters of recommendation, and years
of outstanding classroom performance. Since the 1920s, and

especially in the years after World War II, the SAT has grown in importance as a determining factor in the admissions process.

What is the SAT like? What types of mathematics does it test? A student taking the SAT would encounter several sections testing verbal ability and mathematical ability, along with one experimental section, which is used by the Educational Testing Service (the makers of the SAT and many other exams) to field-test new items. The student has a total of seventy-five minutes to solve sixty mathematics problems. Already it is obvious that time is at a premium: ETS only schedules a working time of a little over a minute per problem. As might be expected, it is difficult to finish this exam. The way the test is scored, though, it is not necessary to answer all items to obtain a good score. The scale for this exam is from a 200 to an 800 with 500 being a good score, 600 very good, and 700 exceptional. Scores on this test are calculated by a corrected score as follows: number of questions answered correctly minus one-fourth of the number of questions answered incorrectly. To receive a score of 500, a student needs a corrected score of about 50 percent, for a 600 a little over 65 percent, and for 700 about 85 percent.

The emphasis, though, is on speed and accuracy. Most of the questions are multiple choice with either four or five options. Test takers cannot argue with an SAT question. They have to learn to think the test makers' way, to avoid traps, and to understand what the test considers important. And what the test makers consider important is not always what is taught in high school mathematics courses.

Since I'm emphasizing word problems, let's look at some example SAT test questions. Consider the following item:[9]

In a race, if Bob's running speed was 4/5 Alice's, and Chris's speed was 3/4 Bob's, then Alice's speed was how many

$B = \frac{4}{5} A$

$C = \frac{3}{4} B$

$C = \frac{3}{4} \cdot \frac{4}{5} A = \frac{3}{5} A$

$A = k \cdot \frac{B+C}{2} = \frac{\frac{4}{5} + \frac{3}{5}}{2} k$

$k = \frac{2}{\frac{7}{5}} , r = \frac{10}{7}$

times the average (arithmetic mean) of the other two
runners' speed?
A. 3/5 B. 7/10 C. 40/31 D. 10/7 E. 5/3

You can already sense the most common complaints high
school students have about this exam: Why? Who would want
to know this? We could ask, common-sensically, how much faster
Alice is than Bob and Chris and then give us some speeds to
work with. This problem, however, emphasizes the test taker's
ability at translation: What do they want me to find, can I avoid
being tripped up by the confusion of the mathematics, and can
I essentially ignore the words to find the basic mathematical
algorithm? Consider another question:[10]

$J = 2P$

$J + 2 = n$

Jim is now twice as old as Polly. In 2 years Jim will be
n years old.
In terms of n, how old will Polly be then?
A. n/2 B. (n/2) + 1 C. (n/2) + 2 D. n + 2 E. 2n

$P + 2 = \frac{n+2}{2}$

$= \frac{n}{2} + 2 = \frac{n+2}{2}$

Here comes this dumb family again. Who would want to
know this? We could ask how much older Jim is than Polly,
but that would be too easy. Once again, you must come to think
the way the SAT thinks in order to even want to attempt this
problem. Also remember: you have one minute to solve it.
    Let's try another one:[11]

$T + J = 750$

$\frac{2}{3} J + J$

Tom and Joe together earn $750 per week. If Tom's salary
is two-thirds of Joe's, what is three-fourths of Tom's
weekly salary?
A. $187.50 B. $225.00 C. $275.00 D. $337.50 E. $375.00

$\frac{2}{3} J = 750$

$J = \frac{750 \times 3}{5} = 450$

$T = \frac{300}{75}$

This one starts out OK. If they had asked for Tom's salary, this might have been reasonable. Then we might have come close to comparing two workers and their respective salaries. But why do we need to know 3/4 of Tom's salary unless they're trying to trick us?

One final example and one of my favorites:[12]

> *Initially, there are exactly 18 bananas on a tree. If one monkey eats 1/3 of the bananas and another monkey eats 1/3 of the bananas that are left, how many bananas are still on the tree?*
>
> *A. 4 B. 6 C. 8 D. 10 E. 16*

Need I comment on this one? Would some math-starved anthropologist sit around counting how many bananas are left on the tree? Or, once again, do the words not matter?

The SAT mathematics test is not entirely word problems, but all questions do emphasize speed and accuracy. The examples of word problems that I have presented, however, could be multiplied endlessly.

What is the Educational Testing Service up to? What is the purpose of the SAT? They publish a little pamphlet, entitled *Taking the SAT*, which describes the function of the test and provides test-taking strategies and sample questions. The pamphlet is distributed free to all high school students and guidance counselors. Describing the mathematics section, the pamphlet says:

> The math questions test your ability to solve problems involving arithmetic, elementary algebra, and geometry. These verbal and mathematical abilities are related to how well you will do academically in college. The SAT does not measure other factors

and abilities—such as creativity, special talents, and motivation—that may also help you do well in college.[13]

Further on, it continues: "Some questions in the mathematical sections of the SAT are like the questions in your math textbooks. Other questions ask you to do original thinking and may not be as familiar to you."[14]

To a certain extent, they are right. They do continue the trend that has traditionally been set down by high school textbooks and focus on abstract mathematical principles or word problems which are essentially meaningless but can be mastered with enough practice. They just carry it to an extreme. A total of sixty questions can determine or influence a student's career. The pressure can easily mount as students' lives get determined by how well they can solve the types of problems listed above. Monkeys could determine your future.

The Educational Testing Service also claims that it does not measure other factors that may contribute to college success, such as creativity and motivation. However, I would disagree to some extent: given the pressure of the SATs in determining college admissions, they have encouraged the creativity and motivation of some students to figure out how to beat the SAT. In other words, any help that can be received to improve scores on the SAT is often gladly paid for by students or their parents.

Test preparation does work! I've taught preparation courses for a number of years now in the Knoxville area and I can say, without reservation, that you can teach a student to take the SAT. Now, the student has to be motivated, i.e., understand that it is important to improve his or her score. This motivation can come from the students themselves or from parents or guidance counselors encouraging them to improve their chances of selecting a good college or university. ETS boldly claims that these types of courses don't work. From the aforementioned pamphlet on special preparations for the SAT we find:

> There is a bewildering array of courses, books, and computer software programs available to help you prepare for the SAT. Some of them do no more than provide the familiarization and practice that is described in the previous section. Others are intended to help you develop your mathematical and verbal skills. These are often called "coaching" courses and we are often asked whether they work. Some students may improve their scores by taking these courses, while others may not. Unfortunately, despite decades of research, it is still not possible to predict ahead of time who will improve, and by how much, and who will not. For that reason, the College Board cannot recommend coaching courses, especially if they cost a lot or require a lot of time and effort that could be spent on schoolwork or other worthwhile activities.[15]

Many students and parents have made their decision: thousands enroll yearly in courses sponsored by Stanley Kaplan or the Princeton Review or in hundreds of smaller-scale special programs like the one I do. To show the ambiguity of ETS's position, one of the best-selling study guides for the SAT is published by ETS itself, entitled *10 SATS*, from which the previous examples were drawn!

How would I coach students to take the SAT mathematics section and receive a high score? I would first give them a review of high school arithmetic, algebra, and geometry. Then I would show them how ETS devises questions and what types to expect, especially in terms of algebra word problems. Finally, I would have the students practice and get used to the test and time limits, using ETS's material. I have experienced average gains of well over 100 points on the math sections and sometimes as much on the verbal parts of the test.

At this point, readers might want to ask, So what's the big deal? So what if you can improve scores? Well, the point becomes this: the test is more related to motivation and money; whoever has access to both gets the better colleges. Furthermore, since the SAT is not directly related to high school coursework anyway, it departs from what should be taught in the high school classroom.

The fact that scores can be increased on the SAT is still not my major problem with this exam. More importantly, the test contributes to two attitudes about mathematics which I feel serve to devastate student and public interest in mathematics in general. First, as mentioned earlier, it perpetuates the notion that to do well in mathematics is to work quickly, get the right answer, and ignore all complexities or real world applications. I have already voiced my complaints about this matter.

Second, it makes most people feel dumb. And I don't just mean dumb in mathematics, I mean dumb in everything. You'll recall that ETS claims the SAT is "related to how well you will do academically in college." Well, the proof on this point is far from clear. SAT scores seem to have some modest relationship with grades the first year in college but after that their predictive power breaks down immensely. In chapter 2, I mentioned the Wallach study demonstrating that SAT scores do not seem to predict real-life accomplishments.

Why should all entering freshmen demonstrate a knowledge of mathematics via the SAT? It can only be that we are still in the throes of believing that to do well in mathematics means a likelihood of doing well in other aspects of life. To do poorly in mathematics means that you should expect poor performance in other endeavors you choose. This attitude originated with the inception of the SAT in the 1920s. The first commission, devised to construct the test, stated one of its purposes: The "tests are so constructed that they put as little premium as possible on specific training, and more emphasis on potential promise as distinguished from prior accomplishment." Continuing, the commission said that "a candidate whose educational opportunities have been limited has a much better chance to show his real capacity in a test which is not a measure of specific preparation, and which is devised so that any person may find increasingly harder and harder problems in which to demonstrate his ability."[16]

In other words, your performance on this mathematics test helps us tell how smart you are and how well you'll do in all college courses. This is too strong. The test cannot accomplish this purpose. As Crouse and Trusheim argue, in their book *The Case Against the SAT*, there is evidence to indicate that the SAT may not add any more information that the high school record doesn't already provide in helping admissions officers select students for their school. In other words, do away with the SAT and the same students get in.

But the impact on mathematics education is where the major consequences are felt. Mathematics needs to change. It needs less rote drill, more problem solving, more real life applications, and greater care in problem construction. An attitudinal shift has to take place before mathematics is respected and enjoyed by a great percentage of the populace. The change has to start with the mathematicians and the teachers of mathematics. We'll see in a later chapter some attempts at this change.

And word problems, in their present format, must go. Instead, make them more real, challenging, and interesting, and be prepared to help students work them. My train problem would be as follows:

Two trains leave at the same time from different cities. The problem is to decide when these two trains would meet if they were traveling toward each other. First, consider if this problem is important enough to invest your time and energy in it. Second, decide what information, mathematical or otherwise, you would need to solve this problem. Consult whatever sources are necessary and don't forget to ask questions as you do so. Report back whenever you feel you have accomplished something worthwhile.

Notes

1. Mary P. Dolciani, Robert H. Sorgenfrey, William Wooten, and Robert B. Kane, *Algebra and Trigonometry: Structure and Method* (Boston: Houghton Mifflin, 1977), Book 2, p. 221.

2. Herbert A. Simon, Dan A. Hinsley, and John R. Hayes, "From Words to Equations: Meaning and Representation in Algebra Word Problems." In Herbert A. Simon, *Models of Thought* (New Haven: Yale University Press, 1989), vol. 2, p. 469.

3. Ibid., p. 473.

4. Ibid., p. 471.

5. Richard P. Feynman, *"Surely You're Joking, Mr. Feynman!" Adventures of a Curious Character* (New York: Bantam Books, 1985), p. 266.

6. Ibid., p. 267.

7. Ibid.

8. The SAT has recently changed its name from the Scholastic Aptitude Test to the Scholastic Assessment Test.

9. *10 SATs: Scholastic Aptitude Tests of the College Board* (New York: College Entrance Examination Board, 1983), p. 25.

10. Ibid., p. 34.

11. Ibid., p. 187.

12. Ibid., p. 107.

13. *Taking the SAT 1992-93* (New York: College Entrance Examination Board, 1992), p. 3.

14. Ibid., p. 16.

15. Ibid., p. 4.

16. James Crouse and Dale Trusheim, *The Case Against the SAT* (Chicago: University of Chicago Press, 1988), pp. 23-24.

4

Testosterone Trigonometry

The argument runs as follows: some babies, while in the womb, are exposed to high levels of testosterone. Exposure to the testosterone slows the development of the left hemisphere, diminishing its size and retarding the development of certain verbal abilities. Consequently, the right side of the brain is enlarged along with certain parts of the left hemisphere that surround the testosterone exposed area. Thus, spatial abilities, which depend on the right hemisphere, and *verbal* reasoning abilities, located behind the posterior language area on the left side, become more highly organized and develop more rapidly. Certain side effects are also produced: left-handedness, myopia, and a susceptibility to immune diseases. As the child gets older, the increased capacity for spatial reasoning attracts the child toward the area of mathematics, a discipline highly amenable to spatial abilities. This talent for mathematics will continue to blossom in the child's early years, despite the influence of parents, peers, and environment. It does not matter what courses are taken in school, because the propensity for mathematics will develop despite certain academic prerequisites. The child will continue to develop mathematically, which will lead to a career as a scientist or engineer. Since testosterone is more likely to be found in male births, this is why we have more males excelling in math than females and more men choosing math-related careers.

The preceding line of reasoning is developed in an article by Camilla Persson Benbow entitled "Neuropsychological Perspectives on Mathematical Talent."[1] For over a decade, Benbow, at Iowa State University, along with her colleague Julian Stanley, at Johns Hopkins University, has been trying to present evidence on the genetic nature of superior mathematical talent. Their philosophy is summed up in a quote from the mathematician Krutetskii: "Anyone can become an ordinary mathematician; one must be born an outstanding, talented mathematician."[2] Benbow and Stanley argue not only that superior mathematical talent is not influenced by environmental factors but that such talent is more evident in males than females. Their evidence derives mainly from school children selected for The Study of Mathematically Precocious Youth (SMPY), founded in 1971 by Stanley. Students in SMPY are selected via high scores on the mathematics section of the SAT, administered to them when they are in eighth grade.

In this chapter I will try to cast serious doubt on the arguments of Benbow and Stanley and other associated attempts to link mathematical talent with genetic factors. I will argue that genetic rationales are actually simplistic substitutes for two very complicated questions:

1. Why does someone choose to become a mathematician? and

2. How does a person become good at doing real mathematics?

Unfortunately, genetic simplicities serve mostly to perpetuate existing stereotypes that mathematical reasoning is a masculine quality, and do not help us understand how or why someone becomes an exceptional mathematician.

Are Males Better at Math than Females?

First, we should examine some of the evidence. How do we know that males are better at math than females? Answers to this question are usually drawn from two sources: (1) males score better than females on the mathematics sections of standardized tests, especially the SAT; and (2) more males become mathematicians, scientists, or engineers, thus increasing the chances that they will become good mathematicians.

Scores on the SAT Mathematics test have traditionally been used to chart the discrepancy between male and female performance. Over the past fifteen years, males have consistently outperformed females by about 40 to 50 points on average (remember the SAT Math scale: a minimum of 200, a maximum of 800, with averages around 500). This means that males outperform females on the types of problems that we encountered in chapter 3.

Is there other evidence, besides the SAT, indicating that males are better at math than females? Logically, it would seem to follow that males are better in school math classes than females. However, when we look at grades in mathematics courses, particularly in high school, we find the opposite: girls almost consistently average higher grades than boys. There seems to be no question in this regard. Meredith Kimball,[3] in her review of the literature, amply demonstrates the superior performance of women in mathematics classes. The problem, she notes, is that educators don't take grades seriously, but rather lavish their attention on test scores. As she says "Although there is ample evidence of young women's superior math achievement when grades are used to measure achievement, they have not been considered seriously in the literature on mathematics achievement."[4] For some reason, we still seem to think that only test scores measure real ability, not mere grades. Imagine a young woman in high school who makes straight As but perhaps only average SAT scores. What future plans will she make?

Do all standardized tests indicate the same male superiority? The answer is a resounding no. In an article on "Gender Differences in Mathematics Performance," Janet Hyde, Elizabeth Fennema, and Susan Lamon[5] analyzed 259 separate studies on gender difference, spanning the years 1967 to 1987. These reports ranged from academic studies to results of several major standardized exams, including the SAT, the California Achievement Test, the Graduate Record Exam, the Iowa Test of Basic Skills, and the National Assessment of Educational Progress.* Taken together, these studies reported gender differences on a total of 3,985,682 subjects (1,968,846 males and 2,016,836 females).

Are males superior in mathematics performance? The authors concluded that their review of these research reports did not support this contention; in cases where there were differences the effects were very small. Of the 259 studies, 51 percent indicated superior male performance, 7 percent were exactly zero, and 43 percent reflected superior female performance. But in most cases the differences were small and did not support the general statement that males are better than females at mathematics, as measured by standardized tests.

If one of the standardized exam results were removed, then the differences become even more negligible. Guess which exam?

---

*These tests are designed to measure mathematical and verbal reasoning abilities, along with other skills, in grades K–12.

The SAT, of course. The SAT accounted for 20 percent of all the data and had a disproportionate influence on the results. The authors treated the SAT in a separate section indicating that its gender differences were more pronounced than on any other test or in any other study. They concluded that those students taking the SAT may not be a representative sample of all boys and girls who take mathematics exams. Thus, once again, we are staring directly into the face of the powerful influence of the SAT in determining our perceptions of gender differences.

There were, however, some interesting gender differences recurring throughout the studies reviewed by Hyde, Fenema, and Lamon. Females seem superior in computation aspects of mathematics, i.e., problems involving fixed algorithms, but worse at problem solving. Furthermore, math differences on standardized tests seem to emerge more clearly in the high school years. Whereas until about age fourteen or so boys and girls are about even in math performance, major differences on tests seem to manifest themselves when students start high school. Obviously, these differences could then have a major impact on a student's life. Test scores are determinants of college admissions and thus future career possibilities. We are confronted once again with the dilemma of the female student who makes good grades but doesn't test "well" for whatever reason.

Could the innate gender differences in mathematics performance be showing up on tests but not in the classroom? Is testosterone so selective? Or should environmental factors be considered? As the previous authors concluded, "the gender differences in mathematics performance, even among college students or college-bound students, are at most moderate. Thus, in explaining the lesser presence of women in college-level mathematics courses and in mathematics-related occupations, we must look to other factors, such as internalized belief systems about mathematics, external factors such as sex discrimination in education and in employment, and the mathematics curriculum at the precollege level."[6]

The first major indicator of male superiority, performance on standardized tests, has been shown to be limited mainly to minor differences on the SAT. What about our second major indicator: that women do not choose math careers as often as do men. Throughout the 1960s, '70s, and '80s, males have dominated these technical areas in terms of college and graduate degrees. Do women then not have the talent for becoming scientists? Is "poor" math performance indicative of thought patterns that disincline them from pursuing more logical, analytical types of disciplines?

I think the answer to both questions is no. There are other logical reasons why women do not choose math careers as often as men do. Meredith Kimball, in the article previously mentioned on grades, offers several social reasons that might inhibit women from technical careers. First, she suggests that women may attribute good performance in the classroom to effort rather than ability. In other words, they can achieve with effort but, at heart, they may not believe that they have a true ability for mathematics. Second, sex-role conflict may be important: women may be discouraged from taking mathematics courses; they may not be given help in pursuing math as a career; or their achievement may be trivialized. Perhaps their femininity may be questioned if they pursue a discipline so traditionally male. Even parents may discourage a daughter who is excelling in a nontraditional area.

Women have not chosen to major in mathematics or technical areas. There has been no encouragement, no role models, and no general belief that they could perform the duties associated with these fields. When this attitude changes so will the number of women in technical fields, a fact that is easily demonstrated.

With the concern in the late 1980s over the lack of women in these areas, special programs were designed to recruit females. These programs have increased the number of women in technical occupations dramatically. More women today are becoming scientists and engineers at a faster rate than males. If this continues for another decade, the dominance of males in technical careers may become a moot issue.

Women are becoming mathematicians and engineers at increasing rates; they make better grades in high school math classes; and they test just as well as males, except on the SAT. So what evidence could Benbow and Stanley use to support a biologically based theory of male superiority in mathematics?

The Benbow and Stanley Studies

The question is worth repeating: What evidence could Benbow and Stanley use to support the idea that males are superior to females in mathematical ability and that this difference is biologically based? For almost two decades, talent searches have been conducted at Johns Hopkins University in Baltimore, trying to identify what Benbow and Stanley call Mathematically Precocious Youth. To isolate these individuals, both verbal and mathematics portions of the SAT are given throughout the country, usually to students in seventh and eighth grades. The authors want to select students before they enter high school, to target them for special programs at Johns Hopkins and other universities.

How do these students perform? Of several thousand tested, both males and females seem to score in the same range on the SAT Verbal (in the high 300s out of a possible 800); however, males outperform females on the mathematics portion by 30 to

40 points on average (mid 400s to low 400s out of a possible 800). Furthermore, Benbow and Stanley examine the number of students who score in the top ranges of the test, with scores at or around 600. When examined this way, males outnumber females by 3 or 4 to 1. Since not that many students score in these top levels, this amounts to about 3 or 4 percent males to about 1 percent females in the total sample. Since seventh and eighth graders could not possibly have taken different courses in mathematics (most advanced math courses are given in high school), then, as the authors suggest, these differences must be based on factors other than environmental causes. From these data, Benbow and Stanley conclude that "sex differences in achievement in and attitude toward mathematics result from superior male mathematical ability, which may in turn be related to greater male ability in spatial tasks."[7] And then onward to the testosterone!

So are males smarter than women in mathematics and is this difference linked ultimately to physiological factors? I believe there are a number of problems with the conclusions that Benbow and Stanley draw from their research. These problems make it difficult to support the broad generalizations about male superiority in mathematics; in fact, their research tends to obscure exactly why it is that someone becomes good at a discipline like mathematics.

In the first place, I would argue that Benbow and Stanley are making too much of the small differences in scores on the SAT. Remember that the average difference in the mathematics score for boys and girls in their studies is about 40 points, out of a possible 800 on the test. Even though this difference may be statistically significant, is there a strong enough difference to claim male superiority and then try to link it to physiology? I think not. Other researchers, as noted earlier, have not found dramatic differences in test scores on anything but the SAT. The SAT may differ because the students who take the test may already be a select group. Are Benbow and Stanley preselecting children who, for whatever reason, have begun to show an interest in mathematics? Can we really assume that eighth-grade boys and girls are the "same" when it comes to interests and motivations?

Benbow and Stanley also find a higher percentage of males scoring in the highest ranges on the SAT, about 4 percent males to 1 percent females. Once again, are these differences large enough to claim some type of male superiority? Also, let me apply a little statistics to their data. There is a good chance of the following happening: if there are more males in the top ranges of scores on the SAT, there is also a good chance there are more males in the low ranges on the same test. In other words, by giving a group of students some type of test, we are about as likely to get as many high scorers as low scorers. Unfortunately, Benbow and Stanley do not report their data on this point. Yes, males may outnumber females on the high end but there is a good chance they also outnumber them on the low end. Thus, there may be more "smart" males and more "dumb" males than their corresponding female counterparts.

The problem this raises is obvious: if we are going to tie these results to testosterone effects at some point, then we are going to have to sort the good testosterone from the bad testosterone. It seems that the same testosterone (unless someone derives a math potency rating for sperm) is producing some "smart" boys and some "dumb" boys. At this point, the entire genetic argument should seem somewhat far-fetched. Even if more boys are better at math, does this mean we draw conclusions about factors at birth? More boys are better at basketball, we might say. Is there something genetic to this? More girls wear dresses than boys: should we seek some type of estrogen effects in the womb? As can be seen, the argument starts becoming ridiculous if we stretch it to areas other than ones that are so sacrosanct, such as mathematics. Because we seem to value mathematics so much in this culture, we are willing to endure almost any type of argument to preserve its uniqueness.

The argument for male superiority also takes a beating when we examine certain statistics related to the SAT itself. It has been known for some time that ethnic groups, on average, score differentially on the SAT. The standard finding is as follows: Asian Americans score the highest on the mathematics portion of the SAT, followed by Caucasians, Hispanics, American Indians, and African-Americans. I examined this data and broke it down not only by ethnic group but also by males and females with the results appearing in table 4.[8]

Table 4

SAT Math Averages for 1990 by Sex and Ethnic Group

	Male	Female
ASIAN-AMERICAN	549	506
CAUCASIAN	515	469
OTHER HISPANIC	460	413
AMERICAN INDIAN	458	419
MEXICAN	453	408
PUERTO RICAN	425	388
AFRICAN-AMERICAN	398	375

Males clearly outperform females by about 40 points or so in each ethnic group. The curious point, in my view, occurs when examining, the female averages. Asian-American *females* score higher than all male groups, except for their male Asian-American counterparts and Caucasians. However, the difference between the average for Asian-American females and that for Caucasian males is only 9 points, clearly not very great. This is a powerful argument for cultural differences in mathematics performance. The expectations, the home environment, the schooling possibilities, or other factors are clearly contributing to substantial ethnic differences in mathematics performance. We can only maintain a biologically based theory if we are willing to assume ethnic differences in testosterone—that one group is more potent than another. This is ridiculous.

Even if we ignore all of the above arguments and admit that more seventh- and eighth-grade males are truly superior in mathematics than females, we are still left with a major problem. There is no good evidence that superiority or "precociousness" in mathematics at the seventh or eighth grade will lead someone to become a good mathematician later on in life. Nor is there evidence that they will even choose mathematics as a career, let alone come to master it at a high level. How does someone become a mathematician? What factors lead to mathematicians becoming excellent in their field? These are the truly important questions and ones that Benbow and Stanley fail to address. To seek answers to these questions we must turn our attention elsewhere.

How Does Someone Become an Exceptional Mathematician?

Benjamin Bloom has spent a lifetime trying to understand what produces talented adults. In one of his most ambitious research projects, published in 1985 under the title *Developing Talent in Young People,* Bloom[9] and his associates at the University of Chicago demonstrate how exceedingly arduous the task of nurturing talent can be. They studied adults who have already received acclaim in one of six areas—Olympic swimmers, world-class tennis players, concert pianists, sculptors, research mathematicians, and research neurologists. All those selected for the study were in the top twenty-five in the United States in their category, based on national ranking systems. All participants and their parents were interviewed at length about the details of their childhood: what had happened at various ages, what were the child's interests, what teachers and/or coaches were employed, how did they like school, and so forth.

After years of gathering and analyzing data, Bloom and his associates concluded that, "The study has provided strong evidence that no matter what the initial characteristics (or gifts) of the individuals, unless there is a long and intensive process of encouragement, nurturance, education, and training, the individuals will not attain extreme levels of capability in these particular fields."[10]

I will focus on the research mathematicians, deriving data from the chapter in Bloom's book written by William Austin. The informants in Bloom's study were all recipients of Sloan Foundation Fellowships, a highly competitive award given to mathematicians under the age of forty. In addition, to qualify for the study, these mathematicians had to be cited often in Science Citation Index, a measure of how much your work is being used by colleagues, and they had to be recommended by mathematics professors at top American universities. Twenty people met all the criteria and agreed to participate in the study; parents of seventeen of these were interviewed. Only one of the twenty was female. Is this evidence for biological superiority? Let's examine some of the characteristics of growing up before we reach a conclusion. Curiously, 70 percent were the oldest children in their families, a fact which we will consider later.

The development of these talented individuals was divided into three periods: the early years, the middle years, and the later years. Each period displayed its own characteristics of learning. In the following accounts, both the mathematicians and their parents speak for themselves, in discussing how and why certain developments took place.

What were the parents of these research mathematicians like? Most were well educated: 70 percent of the fathers had advanced college degrees and 55 percent of the mothers had at least one college degree. More importantly, education and achievement were highly valued by the parents. "I think my father has a terribly serious commitment to serious intellectual activity—as being the entire purpose of life, so to speak. . . . His entire existence is devoted to the life of the mind."[11] At the same time, many parents did not try to specifically direct the interests of their children. "I think it's a waste to try to make a child into something you want rather than providing them with the things they are interested in and letting them become what they want to."[12]

Although the parents did not try to push too hard in a specific direction, they did encourage the traits of working hard, doing well, and being precise. "For some reason, I don't know why, I felt it important that I do well. My parents never really explicitly said that, but I was somehow encouraged to try to work hard."[13]

In many cases, the parents provided the role model for what it meant to work hard. One mother recalled:

> My husband spent a lot of time working in his study, reading and writing. . . . He spent almost his whole life in his study. . . . When —— was about four, he started using my typewriter. When he was about six, he told me he was going to write a book. So he typed a book, it was pages and pages of no words, just pretend words with spaces. . . . He was trying to imitate his dad.[14]

Most of the young mathematicians, at an early age, seemed to be curious and inquisitive. These signs of creativity were often nurtured, with parents encouraging and trying to respond to their children's questions.

> He asked a lot of questions . . . he had intellectual talent. . . . I remember he wanted to know how everything worked, he wanted to know about everything.[15]

> The thing I remember most—there were workmen in the home often. He followed them around all the time. I thought he would drive them crazy. He asked them question after question.[16]

> I listened to my child . . . I talked with him . . . I answered his questions, I read to him.[17]

Very often, the mathematicians themselves remembered being inquisitive. "I've always had this urge to figure out anything that I didn't know,"[18] or "I was always interested in how things worked. I would take toys apart and look at the gears and I was fascinated with valves and gauges and dials."[19] But the children were also content to play alone. "He would spend hours building a tower of blocks, precariously balanced. There would be a wail of exasperation and anguish when it finally collapsed. And then he would start redoing it."[20]

Parents would play with the children, especially games.

When we were young, both my brother and I liked to play games. My father was especially good at playing games with us and keeping it fun. I remember him telling me some really exciting things, which would turn out to be some kind of math, and I would play with the ideas. My father was a sensitive teacher, and I'm sure he did all kinds of subtle things that didn't seem like teaching.[21]

At a young age, both parents and some teachers began to encourage the children to pursue answers on their own, to figure things out themselves.

He would find out different things. Do what you call research. We would send for all kinds of pamphlets. After he was interested in fire trucks and trains, he would be interested in planes, and so we would find out about them—the types. I probably got him started. As he got a little older, he would do it on his own.[22]

I have a sense that [my mother] encouraged my curiosity. . . . I was very interested in astronomy, stars, planets, and space stuff. And to answer my questions she would read me the astronomy articles [from the encyclopedia]. So [I certainly learned] all the standard intellectual values: that you find out things by reading, if you want to find something out, you try to look it up in an encyclopedia, and if it's not there, you go to the library.[23]

I would get some crazy idea and [my teacher] would say, "Let's go and do some research." . . . My father and the good teachers would definitely not tell me the answer.[24]

Many parents felt that their children were special, in the sense that they were intelligent and learned quickly. The children picked this up from their parents and felt that they were expected

to do well at the things they tried. In addition, the children were
heavily encouraged to read. They were also supported in doing
scientific or mechanical projects, or in building models of things,
all before the age of twelve.

> I think I spent a lot of time by myself as a youngster. The
> first dollar I ever saved I spent on a model airplane. I sanded
> it, glued it, put it all together and painted it. I just fell in love
> with the whole process. The planes got bigger and bigger. I
> was interested in the way that they looked. I never wanted
> to fly them. I liked them as art objects. I liked them to look
> as close to the real thing as possible. . . .[25]

> We bought him chemistry sets, crystal radio sets, etc., despite
> the fact that a lot of people didn't feel he should have those
> things. But he was interested, and I thought that a child should
> be given the tools to work with what he is interested in.[26]

> When he was seven, —— approached his father with the idea
> of doing ham radio work. He got his license at the age of nine.
> We actually moved to provide him more space for his hobbies.
> He took over the entire basement. He built transmitters and
> receivers, experimented with his chemistry set, and set up a
> darkroom. . . . He was always interested in how appliances
> worked. I sent him to repair shops to learn how these things
> were put together.[27]

Although academic achievement was valued by most of the
parents, the young mathematicians attached very little importance to their experiences in elementary school. Nineteen of the
twenty attended public elementary schools. Most were not very
enthusiastic about their school experiences, although they excelled in most of their subjects. Eight of the mathematicians had
difficulty relating to other students. Most also felt that school

math, at this point, had very little impact on their overall conceptualizations.

> I never really learned any mathematics in school. I've learned everything I know on my own, except some arithmetic. Once I started studying on my own, I never learned any mathematics in school. I've had some inspiring teachers, but they've always been intelligent enough to see that I'm very independent and that the best thing they could do for me was to give me the books and let me work on my own.[28]

The second phase of development spans the years from the start of junior high through graduation from high school. The authors see this phase as transitional, extending the activities and interests of the earlier phase and foreshadowing some of the developments that would occur later in college. Parents continued to be interested in the progress of their children, with actual support ranging from encouragement to arranging special programs. The students at this point continued to "do" things on their own, especially in science.

> He would set his alarm and get up in the middle of the night, get dressed, and go out and see the stars.[29]

> One of my big pleasures was to master something or other, whatever it was. I'd have little projects from time to time, learn to do this and that. And so I've always enjoyed the feeling of accomplishment that you get from learning to do something that is difficult.[30]

Independent learning continued to a great extent in both science and mathematics during these years.

Once I got to high school I spent enormous amounts of time just by myself doing mathematics or analyzing board games—working out strategies and trying to analyze how to play a game. . . . And I read. I taught myself some calculus, using texts that my sister had from college.[31]

No one really checked my work, no one. I knew if it was right. I knew because there were answers in the back of the books and things like that. You don't have to have someone check your work. You know if you are doing it right. I suppose I made some mistakes, three or four a year maybe. But you know if you are right, it makes sense.[32]

In high school, mathematics and science courses were easy for these students and many excelled in other areas. They seemed to be divided on their perception of the high school years, with half viewing them positively and the other half negatively. What impressed almost all of them, however, were teachers who "knew their subjects" or were "interested" in what they were doing.

My geometry teacher had quite a positive effect. She liked geometry and she realized it involved thinking. This teacher actually understood what she was saying and she would try and get the class to understand. I think she was a very traditional teacher. Her approach was completely traditional, but the point was that she obviously enjoyed the subject and she knew what she was talking about. . . . If you want a positive experience, you need to have an interpersonal type of relationship with someone who succeeds in creating a positive experience. My view is that although the materials certainly have some significance—it's better to have a good book than a bad book and it's better to have this method than that method—that sort of second-order effect, really, it doesn't tell the tale. A good educational experience has to do almost entirely with interpersonal effects. . . . The point is that you tune in to the person. The main function of teachers is to make things interesting, to produce positive motivation, to serve as an example, do something that you can copy to a certain extent.[33]

The final stage of development, the later years, extends from college through graduate school. In this phase, the mathematicians learn to do "real" mathematics, which includes doing research and aiming themselves toward a career in mathematics. The early years and the middle years produced a sense that they were inclined toward mathematics and sciences, that they were smart, and that they could do independent work. Parents expected them to go to college. Less than half, though, actually thought about majoring in mathematics in college. Despite their good experiences and abilities in mathematics at earlier ages, most still did not see themselves as "mathematicians."

I went to college with the intention of becoming a chemistry major. I also had a very good math professor. He was a good lecturer. He had a little seminar for people who wanted to learn a little extra on the side, and I joined that. He was willing to talk with me. I saw him often. [My choice of math as a major] may have been [a result of] the two professors I had, the chemistry professor was very distant.[34]

By their sophomore year, however, many of these mathematicians had done well in math classes, had come to the attention of their teachers, and had begun taking honors classes. This point marks the emergence of their desire to be mathematicians.

The teacher that I had for freshman calculus was picked out to start an experimental program in mathematics in which essentially all of undergraduate mathematics was going to be shoved down our throats in something like three semesters. And then we would start doing graduate work. I was one of five or six students picked out to do this, on the basis of how well I had done in the first semester, presumably on the basis of being able to understand theoretical material. He had personally picked us out. Most of us went on to become professional mathematicians.[35]

Teachers in this later period seemed to have even more of an impact than they did in earlier years, as role models and as people doing important work in mathematics.

There was one teacher, I was completely in awe of him. I liked his style. It just appealed to me, it was just right. You can tell. I think the way you learn mathematics is, in some sense, you imitate. You don't know. You do it. So I was around him and I would imitate. He'd approach a problem a certain way. And that's what really it's about. How you go about it. And this is the most important thing of all: taste. It's something enormously important in mathematics. There are very smart people who do very hard things, extremely hard things. But these things are uninteresting. They don't have the taste to really pick the gems. He had good taste.[36]

The work in mathematics starts out as fun, and, one hopes, it continues in that vein. At some point in college, though, a certain commitment to mathematics as a career begins to fashion itself.

I can tell you quite frankly when I first decided I'd stop playing and really work in mathematics. I was about second or third year in college. I'd been playing at it. I'd learned. I'd read and

all this, but that's play, it's passive. And then I did absolutely
nothing else [but mathematics] for many, many, many years.
It's impossible. I don't know anybody who does first-class
mathematics who doesn't work all of the time. Your world
becomes very small.[37]

Going to graduate school in mathematics involved a more
conscious choice of the more prestigious graduate programs and
determining which mathematicians were at the top schools.

I think most people working mathematics have, at some point,
close contact with a leading center of mathematics. In fact, I
can't think of any example where a person has done great work
without having such contacts.[38]

Even after finishing a doctorate, however, there was still some
sense of development, that a career and the possibility of doing
original work still had to be worked out.

I really had grave doubts about my ability to do creative
mathematics. There's no way of knowing until you do it. You
just can't tell until you get there. It turned out somewhat to
my surprise, that I was able to do research, and to my even
greater astonishment that I was pretty good at it.[39]
    You can work for weeks and be absolutely nowhere. But
it's necessary, the time is not wasted, because if you hadn't
done the work and gotten nowhere, then you wouldn't get the
idea that may solve the problem. It's half hard work and it's
half play, but mostly it's just engrossing. You have to get to
the end of this so you can sleep again. In some sense solutions
are unpredictable. The only thing that you can predict is that
you won't find solutions if you don't work.[40]
    When I went to graduate school, I decided, "Okay, this is

it. I'm going to spend a few years of my life really deeply learning something." And at the end of four years I knew a lot of mathematics and I started to do research. I got involved in my profession and continued to learn and learn. And at the end of a decade of that there's just no comparison. You need that kind of time. Nobody really does good mathematics of a certain kind without having put in a lot of time.[41]

How Are We to Nurture Talent?

Certain elements in the talent development process must be attributed to chance. Why do children play at this game or that? How many facilities are available with which the child can play? Why do parents encourage one thing and not another? Other elements, however, are more deliberate: that the parents themselves work hard and expect the best from themselves and their children; that they support their child's interests without directing them too much or expecting the child's interests to gratify their own desires; and that the child be continually encouraged to learn, to play, and to ask questions.

The length of involvement must not be underestimated. Bloom's portraits indicate that development is taking place over a dozen years or more, with many failures and some small successes. Parents must realize that it takes a long time for children to develop adult abilities, even if they seem precocious at an early age. As Bloom comments, "Only a few of these individuals were regarded as child prodigies by teachers, parents, or experts. Even those few who were regarded as unusual at an early age were not perceived to be able to do things in their talent fields that would compare with mature talented persons in these fields."[42] Practically none of the mathematicians had reached mature levels by the time of seventh and eighth grade. "Precociousness in a talent field is not to be dismissed, but it can only be realistically viewed as an early stage in talent development."[43]

Will a test, such as the SAT, identify those who will eventually become talented mathematicians? "Even in retrospect, we do not believe that the perfecting of aptitude tests or other predictive instruments would enable us or other workers in the field to predict high-level potential talent at these early ages."[44] This includes possible biological predictions; for, if we don't know if someone is going to be a superb mathematician by watching their talent at the ages of eleven or twelve, how could we possibly know by simply ascertaining their sex or the traits of their parents? It's just not that simple!

How, then, are we to account for sex differences? Why are more males, even in Bloom's sample, exceptional mathematicians? I think the answer is clearer if we realize the long process of development that is needed to inculcate certain values and attitudes in children. Parents must be hard working, expect the same from their children, encourage intellectual pursuits, and help children ask and answer questions. Parents must help girls as well as boys think about and play with the materials of mathematics and science. The toy rockets, airplanes, the chemistry sets, and gadgets, not to mention the computers that seem so male-oriented have to be seen as nongender specific. Boys and girls can play at these things and parents will encourage them to do so. Physiology is not nearly as important as possibility, the attitude that anyone can become good, under the right circumstances. Only if we want females to do well in mathematics will they succeed.

Notes

1. Camilla Persson Benbow, "Neuropsychological Perspectives on Mathematical Talent." In *The Exceptional Brain: Neuropsychology of Talent and Special Abilities,* ed. by Loraine K. Obler and Deborah Fein (New York: Guilford Press, 1988), pp. 48–69.

2. Ibid. p. 52.

3. Meredith M. Kimball, "A New Perspective on Women's Math Achievement," *Psychological Bulletin* 105 (1989): 198–214.

4. Ibid., p. 203.

5. Janet Shibley Hyde, Elizabeth Fennema, and Susan J. Lamon, "Gender Differences in Mathematics Performance: A Meta-Analysis," *Psychological Bulletin* 107 (1990): 139–55.

6. Ibid., p. 151.

7. Camilla Persson Benbow and Julian C. Stanley, "Sex Differences in Mathematical Ability: Fact or Artifact?" *Science* 210 (1980): 1264.

8. Data obtained from the College Board.

9. Benjamin S. Bloom, ed., *Developing Talent in Young People* (New York: Ballantine Books, 1985).

10. Ibid., p. 3.

11. Ibid., p. 272.

12. Ibid., p. 273.

13. Ibid., p. 275.

14. Ibid.

15. Ibid., p. 277.

16. Ibid.

17. Ibid.

18. Ibid., p. 278.

19. Ibid.

20. Ibid., p. 279.

21. Ibid., p. 281.

22. Ibid., p. 283.

23. Ibid.

24. Ibid.

25. Ibid., p. 287.
26. Ibid., p. 288.
27. Ibid.
28. Ibid., pp. 292–93.
29. Ibid., p. 298.
30. Ibid.
31. Ibid., p. 300.
32. Ibid., p. 301.
33. Ibid., p. 307.
34. Ibid., p. 316.
35. Ibid.
36. Ibid., pp. 319–20.
37. Ibid., p. 321.
38. Ibid., p. 323.
39. Ibid., p. 326.
40. Ibid., p. 327.
41. Ibid., p. 328.
42. Ibid., p. 533.
43. Ibid., p. 538.
44. Ibid., p. 533.

5

Last Place Blues

In the State of the Union address for 1990, then President George Bush proclaimed as one of his educational reform goals that American students, by the year 2000, should rank first in the world in math and science. Basing his statistics on international surveys that show the United States invariably in last place in math and science comparisons, Bush prophesied his panacea: "Real improvement in our schools is not simply a matter of spending more, it is a matter of expecting more."[1] President Clinton has continued to support this national goal. How do we know that the United States is in last place in math and science? What would it mean to try to make us first? How much "more" should we expect? Why are we even worrying about this problem in the first place? By examining the studies and the rhetoric behind the current math and science debates, I hope to show that our goals are misguided: trying to make our students "first" will do very little to change the perceived problems in American society and may, in fact, cause more harm than good.

To see how such a nonintuitive conclusion can be supported,

we must first investigate why it is thought our students are *not* number one in the world. Then we can determine if making them number one will, in effect, help anyone: our students, our schools, our economy, or our president.

A Look at International Comparisons

To begin, it is important to examine the details of the studies of international assessment, which claim to find the United States to be in last place in math and science. A widely cited study appeared in January 1989, under the title *A World of Differences: An International Assessment of Mathematics and Science.*[2] The study was conducted by the Educational Testing Service (ETS)—makers of the Scholastic Assessment Test (SAT)—using methods and materials that had already been developed for the National Assessment of Educational Progress (NAEP), the organization responsible for *The Nation's Report Card*. NAEP has for the last twenty years informed Americans biennially on their progress, or lack thereof, in the basic areas of math, science, reading, writing, and, most recently, history and geography. These report cards have shown a decline in academic performance in these areas in the 1980s for most of the age groups they sampled.

This international study, which was acronized as IAEP (International Assessment of Educational Progress), involved five countries and four Canadian provinces: Ireland, Korea, Spain, United Kingdom, and the United States with Canada adding the provinces of British Columbia, New Brunswick, Ontario, and Quebec, with the latter three testing in both English and French. A representative sample of thirteen-year-olds was drawn at random from each country; each student took a forty-five-minute mathematics assessment consisting of sixty-three questions and a forty-five-minute science assessment composed of sixty questions. Questions were translated into the language appropriate for each country. Why thirteen-year-olds? Many countries stop requiring formal schooling around this age; thus, to insure a sample of children still in school and who have recently had mathematics training, this age was chosen.

Since the results were nearly identical for both the math and science assessments, I will focus on the mathematics test. How did the United States do in this study? On a 0 to 1000 point scale, the United States finished dead last with an average of 473.9; Korea headed the field with an average of 567.8. What is producing the differences? Why are our students so bad? To the credit of IAEP, the study does attempt to pinpoint where differences in performance occur. As is typical of most of *The Nation's Report Card* studies, students' abilities are grouped into a series of scales or levels of performance, with each succeeding level demanding more in terms of mathematical proficiency.

Figure 2 presents descriptions of the five levels of mathematics performance and an example of a problem at each level. The levels progress from basic addition and subtraction problems, through two-step problems involving averages and some algebra, to geometry, and problems involving complex tables and graphs. For the benefit of the curious, try these problems. (The answers are to be found later in this chapter.)[3] The problems reflect material that students should have encountered in an elementary and secondary school curriculum with the emphasis on arithmetic, algebra, geometry, and word problems. These test items also strongly reflect ETS's measurement bias of multiple choice items or simple fill-in-the-blank responses. More complicated mathematics performance items are not included as part of this study.

Where do American students particularly fail to make the grade? Table 5 presents the percentages of students performing

LEVEL   **Perform Simple Addition and Subtraction**

300   Students at this level can add and subtract two-digit numbers without regrouping and solve simple number sentences involving these operations.

LEVEL   **Use Basic Operations
to Solve Simple Problems**

400   Students at this level can select appropriate basic operations (addition, subtraction, multiplication, and division) needed to
solve simple one-step problems. They are capable of evaluating simple
expressions by substitution and solving number sentences. They can
locate numbers on a number line and understand the most basic
concepts of logic, percent, estimation, and geometry.

LEVEL   **Use Intermediate Level Mathematics Skills
to Solve Two-Step Problems**

500   Students at this level show growth in all mathematics topics
in the assessment. They demonstrate an understanding of the concept
of order, place value, and the meaning of remainder in division; they
know some properties of odd and even numbers and of zero; and they
can apply elementary concepts of ratio and proportion. They can use
negative and decimal numbers; make simple conversions involving
fractions, decimals, and percents; and can compute averages. Students
can use these skills to solve problems requiring two or more steps and
can represent unknown quantities with expressions involving variables. Students can measure length, apply scales, identify geometric
figures, calculate areas of rectangles, and are able to use information
obtained from charts, graphs, and tables.

LEVEL   **Understand Measurement and Geometry
Concepts and Solve More Complex Problems**

600   Students at this level know how to multiply fractions and
decimals and are able to use a range of procedures to solve more
complex problems. Students demonstrate an increased understanding
of measurement and geometry concepts. They can measure angles
found in simple figures, understand various characteristics of circles
and triangles, can find perimeters and areas, and calculate and compare volumes of rectangular solids. Students are also able to recognize
and extend number patterns.

LEVEL   **Understand and Apply More
Advanced Mathematical Concepts**

700   Students at this level have the ability to deal with properties
of the arithmetic mean and can use data from a complex table to solve
problems. They demonstrate an increasing ability to apply school-
based skills to out-of-school situations and problems.

$$29 = \square + 16$$

What number should go in the box to make the number
sentence above TRUE?

ANSWER _____

What number does ▼ point to?

① 1

② 2

③ 3

④ 4

Here are the ages of five children:

13, 8, 6, 4, 4

What is the average age of these children?

① 4

② 6

③ 7

④ 8

⑤ 9

⑥ 13

⑦ I don't know.

The length of a side of this square is 6. What is the radius of the circle?

① 2   ② 3   ③ 4   ④ 6   ⑤ 8   ⑥ 9   ⑦ I don't know.

NUTRITIVE VALUE OF CERTAIN FOODS

	Measure	Calories	Protein (grams)	Carbohydrates (grams)
Banana, raw	1	100	1	26
Beef hamburger	3 oz.	245	21	0
Whole milk	1 cup	160	9	12
Doughnut	1	125	1	16
Eggs, boiled	2 eggs	160	13	1

According to the table, what is the total amount of protein contained in
two boiled eggs and one-half cup of whole milk?

ANSWER _____

Figure 2: Levels of Mathematics Proficiency

at each mathematics scale level for each country. As can be seen, performance is fairly high on level 300 (Add and Subtract) and level 400 (Simple Problems) for all countries; conversely, all countries perform poorly on level 700 (Interpret Data) and don't do so well on level 600 (Understand Concepts) either, except for Korea's 40 percent on this level. Thus, we might conclude that thirteen-year-olds, regardless of country, can do simple mathematics problems but strike out when it comes to problems involving geometry or more advanced mathematical concepts. Level 500 (Two-Step Problems) seems to be the main discriminator or the level with the most variation among countries (see Table 5).

What difference does this make? So what if our thirteen-year-olds are only half as good at two-step mathematics problems as Korean students? The report offers some suggestive conclusions:

Table 5

Percentages Performing At or Above Each Level of the Mathematics Scale, Age 13*

	Add and Subtract	Simple Problems	Two-Step Problems	Understand Concepts	Interpret Data
LEVEL	300	400	500	600	700
KOREA	100	95	78	40	5
QUEBEC (FRENCH)	100	97	73	22	2
BRITISH COLUMBIA	100	95	69	24	2
QUEBEC (ENGLISH)	100	97	67	20	1
NEW BRUNSWICK (ENGLISH)	100	95	65	18	1
ONTARIO (ENGLISH)	99	92	58	16	1
NEW BRUNSWICK (FRENCH)	100	95	58	12	<1
SPAIN	99	91	57	14	1
UNITED KINGDOM	98	87	55	18	2
IRELAND	98	86	55	14	<1
ONTARIO (FRENCH)	99	85	40	7	0
UNITED STATES	97	78	40	9	1

*Jackknifed standard errors for percentages range from less that .1 to 2.4.

Many 13-year-olds are within one or two years of completing their study of mathematics. While it can be comforting to learn that almost 100 percent of a country's 13-year-olds have mastered basic addition and subtraction skills, it may be of concern that only 40 percent can use fractions, decimals, and percents, since the 60 percent of students who have not yet developed these skills may experience difficulty with secondary-school mathematics and, if their competence is not increased, may face serious problems dealing with the everyday quantitative problems that confront modern adults. . . .

All of the countries involved in the assessment share common goals for an improved quality of life for their citizens and for successful economic achievements in the world arena. Each society is experiencing rapid technological change that often translates into the need for employees who are better trained in mathematics and science. The 13-year-olds of 1988 will be the 18-year-olds of 1993. . . . If more than 75 percent of the 13-year-olds in a country are competent in intermediate mathematics skills, does that country have a significant social or economic advantage over a country in which only 40 percent of this age group has attained this proficiency level?[4]

This is an extremely good question. Do our thirteen-year-olds affect our economic prosperity? Have they affected it in the past? What factors do provide social and economic advantages? Why, for instance, has Japan been able to advance so quickly and efficiently on economic fronts? Is Japan's improvement tied to the education its children receive? These are complicated questions; for our purposes now we need to consider how the math skills of thirteen-year-olds help us answer these questions.

I will argue that the math skills of our thirteen-year-olds do not help us to understand the social and economic problems that we are facing; nor will enhancing these skills necessarily make us any more competitive in the world economic arena, especially in scientific and technological areas. In fact, attempts to raise math scores, without qualifying certain underlying assumptions, may in fact produce even more children eventually to become adults who will hate math in everyday life.

Aside from three general criticisms, I do not want to fault the methodology of the IAEP study. Within its limits, it is a very professional study, well conceived and executed and psychometrically very sophisticated. I will mainly challenge some of the conclusions the study makes from the data and more importantly the uses to which the study has been put by the president, Congress, and the general public.

First, the three complaints. Since the study tested only thirteen-year-olds, we need to ask how representative are these adolescents of the mathematical abilities of teenagers in general? This particular age was chosen, or determined, by the educational curriculum of other countries, this being the age when mandatory schooling stops. For the United States, however, this is the beginning of high school,* a time when most students are being introduced to algebra and geometry. So are the U.S. students' failures on two-step problems and geometry the results of ability or simply class scheduling? Even IAEP is unsure on this point. After warning of impending social and economic disasters, the report adds: "From another perspective, school administrators may need to question whether poor performance of thirteen-year-olds reflects a generally inadequate program or simply a situation in which concepts have not yet been taught because of the sequence of the curriculum. Performance results may signal the

---

*Students in the United States must attend school until the age of sixteen.

need to reconsider specific syllabi or broad educational standards."[5] If we want students to beat the world by the time they are thirteen, the implications are obvious: have students learn more algebra and geometry by this age by increasing requirements in elementary and junior high schools. But aren't there other consequences of forcing so much math on children at such an early age?

Second, there is the problem of motivation. Students take these tests without any incentive. They are taken from the classroom, brought into the library or cafeteria, and basically told to do these tests. School officials, i.e., principals or superintendents, usually decide if their school will be part of the project. If this is true for every country, then what's the big deal? Well, the value of schooling and performance on random tests may be culturally specific. Do students in Korea do what they're told better than students in America? I don't know the answer but I do propose an interesting experiment: Let IAEP pay the students the next time around for their performance. With this as an incentive would we solve America's math crisis overnight? Would our kids perform better if they were being paid, a value which most American students recognize?

Finally, consider the nature of the test. Like so many standardized math tests, this exam measures performance on simple math items or word problems, questions that may lack real value in the everyday world. The study claims that students need these skills for the work world when they grow up; furthermore, it warns that most students are within one to two years of finishing their mathematics education. If the latter point is true, then every country needs to have an anxiety attack: all students performed miserably on Level 700—understanding and applying math concepts to out-of-school situations. Will another one or two years of schooling raise performance in this vital area enough to produce a technologically skilled worker? I would say no; skills which actually get used in jobs may come from someplace else.

So why is everyone getting excited about slight differences among thirteen-year-olds? The report clues us into its major assumption:

> Thirteen-year-olds who have mastered the skills reflected in the descriptions of Levels 600 and 700 probably represent the pool from which most of tomorrow's mathematicians, engineers, and scientists will emerge. Do these results predict that certain populations will be responsible for a majority of the important achievements in these fields during the twenty-first century? Obviously, the answer to this question depends on the opportunities presented in each society and the support available to young people in each country to pursue and develop their interests.[6]

Choosing to Become a Scientist or Mathematician

What are the opportunities to pursue and develop interests? How do people decide to become mathematicians, engineers, and scientists? Will raising the mathematical ability of thirteen-year-olds mean that more of them will choose to pursue technical fields? These are the questions that address the main assumptions of most international studies and the only ones of practical value.

It is important in this argument not to confuse the past with the future. In the past, the United States has declined economically in areas in which other countries, especially Japan, have progressed rapidly. In the past, the United States has had certain problems keeping up with the economic development of the Japanese, particularly both Japan's incredible ability to perfect and market simple and advanced electronic, computing, and technical devices. They have been masters at marshalling resources to achieve economic gains for their nation. Consequently, the United States world share in these areas—electronics, automobiles, steel, to name a few—has slipped dramatically. To play the devil's advocate for a moment, can we say that our present thirteen-year-olds are responsible for us becoming a debtor nation? This is obviously ludicrous and no one would argue this point. Well, are the thirteen-year-olds of ten years ago, now twenty-three, to blame? This age group would have just barely finished college; to become scientists and engineers they probably have only started their masters or Ph.D. programs with several more years of work before they make an impact. Are the thirteen-year-olds of twenty years ago, now thirty-three, responsible? This group has had time to become technologically sophisticated, to pursue advanced degrees, and to aid in America's technical progress. The only problem with blaming it on this group, children of 1970, is that most of the NAEP comparisons start in the 1970s. The groups that do poorly today do so when compared with the high performance of the children of the early 1970s! These age groups, then, are clearly not responsible for the problems or mistakes of the recent past. In one sense, then, it is hard to blame any of our economic woes on the children of the past twenty years, for they have not been in any position to affect our economic and technological status.

Then we're only worried about the future, about catching up, and correcting whatever mistakes have been made. We need more scientists and engineers to offset the losses that somebody is responsible for. This returns us to our primary question and the one that underlies *The World of Differences*: how does someone select a career as a mathematician, scientist or engineer?

One obvious point to start is in the classroom. Do students

who are really good in math choose careers in mathematics, engineering, and science? Are the students who make the best grades, the As and A+s, the ones likely to continue to pursue this subject? In other words, if we raised the scores of students and made them better in math, would they choose technical careers? I did some checking of this hypothesis: that better-than-average students in mathematics choose careers, let's say, in engineering.

To be sure, engineering programs might preselect people on the basis of grades in high school or grades in mathematics, more specifically. But evidence, I believe, does not support the latter contention: just because people make good grades in mathematics courses does not mean they will be attracted to a career in math or science or engineering. The talk about the grade inflation of the last ten to fifteen years has occurred at the same time that students are choosing not to pursue technical careers as much as they once did.

So, why does someone become a mathematician or an engineer? Why do people choose to pursue these professions? I believe that most discussions of making the United States number one in the world in math and science would be much better spent by trying to answer these questions. Instead of wanting everyone to take more science and math courses in elementary and secondary school, reformers' focus should be centered on trying to understand an important psychological consideration: motivation. What motivates a person to want to do something?

If this question is asked, one answer becomes immediately clear: High grades do not seem to be the motivator. Studies clearly show that females have higher grades, on average, in high school mathematics classes than males. At the same time, however, females choose scientific and technical careers far less often than do their male counterparts. So much for just making an A in the course!

So, what are the motivations for majoring in mathematics and science? In a study for the National Science Foundation, William LeBold and Kathryn Linden[7] examined the career development of 2,835 graduate engineers and 980 student engineers. These engineers were asked to rate how important various factors were in influencing their decisions to pursue an engineering career.

The major factors that influence someone to choose a career in engineering are mostly intrinsic; i.e., engineers reported that they "liked problem solving," wanted a "challenge," "creativity," or "independence." Extrinsic factors, such as the "salary" or "prestige" of the engineering profession were also highly important. Various people had strong influences on them, in particular their fathers, to some extent their mothers, and also other male engineers. High school science and math courses were also rated highly, along with certain hobbies they were engaged in, particularly using a computer. Taken together, these results suggest that majoring in engineering starts as an intrinsic interest and then becomes reinforced by school and work considerations as students become older. For schools to help promote the development of these careers, then, it becomes extremely important to recognize and encourage these intrinsic interests.

Once a person has, for whatever reason, developed a personal interest in a subject like math or science, then the problem for the schools is in trying not to hinder it. Is it possible that students choose technical careers despite their poor performance in mathematics courses? One of the more reasonable recent studies to discuss honestly the value of mathematics is the report by the National Research Council entitled *Everybody Counts: A Report to the Nation on the Future of Mathematics Education*.[8] The report presents evidence that half of the students drop out of the mathematics pipeline, i.e., stop taking courses in mathematics, every year. The report is honest in its appraisal of the causes: "More than any other subject, mathematics filters students out of programs leading to scientific and professional careers. From high school through graduate school, the half-life of students in the mathematics pipeline is about one year; on average, we lose half the students from mathematics each year, although various requirements hold some students in class temporarily for an extra term or a year. Mathematics is the worst curricular villain in driving students to failure in school. When mathematics acts as a filter, it not only filters students out of careers, but frequently out of school itself."[9] They present further evidence that the number of intended mathematics majors among top high school seniors has been dropping dramatically since the mid-1970s.

Although there are many forces behind students dropping out of this pipeline, the report emphasizes that undergraduate mathematics, the mathematics of the college experience, is vitally important, perhaps even more so than elementary or secondary school mathematics. This is especially true since a college degree is necessary for most of the technical careers that seem to be in short supply.

> Undergraduate mathematics is the linchpin for revitalization of mathematics education. Not only do all the sciences depend on strong undergraduate mathematics, but also all students who prepare to teach mathematics acquire attitudes about mathematics, styles of teaching, and knowledge of content from their undergraduate experience. No reform of mathematics education is possible unless it begins with revitalization of undergraduate mathematics in both curriculum and teaching style.[10]

Undergraduate mathematics is primarily calculus. Most reforms of mathematics education focus on high school material; higher education often escapes criticism unscathed. The problem, however, of why students don't choose technical careers may be at the college level and not at the high school level. As the report continues:

> The quality of calculus instruction is a barometer of mathematics education. Since preparation for calculus has been the organizing principle of high school mathematics, calculus receives the inheritance of school practice. Changes in calculus reverberate throughout secondary school curricula, just as changes in school mathematics are magnified by the challenge of calculus. Although other courses need improving as much as calculus, and although many courses are as important or as practical, the unique position of calculus as the gateway from school to college mathematics imposes on it a special burden to be attractive, compelling, and intellectually stimulating.[11]

But is college calculus so organized? "Unfortunately, calculus as presently taught has little in common with the way calculus is used. Many students who enroll never complete the course; many of those who do finish learn little beyond a series of memorized techniques now more commonly performed by computers."[12]

The previous discussion highlights some of the reasons why students might choose scientific and technical careers and why they might not choose them. Mathematics courses, particularly calculus, can serve as a filter to weed out many who might be interested in careers such as engineering, for example. Or they pursue their technical interests despite their mathematical training. But, on the more positive side, perhaps we can learn more about the lack of people choosing science and math careers by asking what careers college students are seeking. What do they want to do with their lives?

At What Do We Excel?

An examination of the percentage of college students receiving bachelor's degrees from 1972 to the present reveals some interesting trends.[13] The percentage of students majoring in subjects centered around business and management, and communications (with the allied fields of broadcasting, advertising and entertainment), has increased dramatically. Computer and information science degrees have mushroomed, increasing twentyfold from the early 1970s. Mathematics has dropped precipitously but so has a field like education. Engineering, curiously, has shown a moderate rise, increasing twofold. Now, on one hand, some might say that students are taking the easier courses, or subjects more centered on materialistic goals and avoiding the tougher academic areas of math and science. (Although, how do we explain the explosion in computer and information science areas? Why do we never hear about the growth in this subject? Why only math and science?) I would propose another explanation. Take the following test: Name the areas in which the United States seems to be unequivocally in first place in the world. At which areas do we all agree that America excels or is the best? Perhaps it comes as a surprise to see that what we are best at—entertainment, advertising, business, communications, and computer software—is exactly what students want to major in.

In other words, our students, if we'd only listen to them, are telling us our strengths and weaknesses. They're telling us that America excels in areas of business and communications and that we're not as good in other technical areas. We can beat the world at putting quality programs on the TV screen but Japan can make a better and cheaper TV set. We are experts at business consulting, high finance, and international marketing but who has the most loose cash to spend on these concerns? We are extremely good at creating computer programs but Japan can build better and cheaper semiconductors. What conclusions should we draw from this?

At one level, we should stop telling our children they're dumb because they seem to score low on math and science tests and start telling them they're smart for investing their lives in areas they know have futures, and exceptional futures at that. I would love to see international comparisons in such areas as creativity, business, advertising, marketing, and perhaps even rock 'n' roll and video games. Who would lead the world then?

At another level, we should pay attention to their choices. For by their choices they are telling us what has happened. They are saying that science and math careers are not attractive, because our leaders of government and industry no longer see them as promising. This is despite all the rhetoric and propaganda surrounding the current cries for educational reform in math and science. If you don't think this is true, consider some of the following items from our treatment of technical aspects of American life in the past twenty years:

1. American automobiles, once the pride of the world, have become inferior in quality to cars produced by both Japan and Germany during the 1970s and 1980s. American technical expertise declined in this area until the 1990s.

2. The American electronics industry has also fallen behind the Japanese in the quality of product made. Recent efforts to revive this industry, by investing in High Definition Television (HDTV), have been met with congressional setbacks in funding.

3. Investments in computer research and development, and in both semiconductor and supercomputer research, have declined dramatically during the 1980s. Steve Jobs (founder of Apple computers) comments on the shortsightedness in developing the computer industry:

If we want to be the preeminent supplier of computer technology, we have to ask what we must do to regain supremacy in semi-conductors, storage devices like disk drives, and video display technology. These are the three biggest cost items in computers and they are also the three biggest chances to do revolutionary leaps forward. If you do not control any of these technologies, you are going to lose. You will probably lose if you control only one of them. We need to be leaders in all three. The problem is that nobody is even asking the questions, much less coming up with a coherent set of strategies to execute the answers. Getting rid of General Noriega is important, but I wish the computer industry would get a tenth of the space on our national agenda that he has. We have to make these issues national priorities. Let's have President Bush start talking about what it will take to win the high-definition video display race. You'd be surprised at how fast things would happen. For example, a venture capitalist would start to write checks.[14]

Lee Iacocca foresees some of the same problems in the auto industry:

American business has got to perform differently in the 1990s. The 1980s were a time of quick bucks, greed, and a lot of corruption. . . . We can't sit around and commiserate with one another—we've got to get good, we've got to compete, we've got to be worldclass. We can't just shout it, we've got to be it. . . . So much of our brain power has been siphoned off into the defense establishment. If you look at it in terms of total R & D this country is terrific, but if you take out defense, there's none left. That's when we look bad against Japan. The military-industrial complex, which I've often believed exists, is not going away. But as you start to take it down, you can free up some of the those brains to do commercial work. Instead of building black boxes and missile silos, they'll be working on instrument panels.[15]

The list could continue; it points out, however, the real problem. American leadership has grown tired of high-tech adventures or has become poor in the management of the resources in these areas. The problem with math and science is not with the children but with the adults, the leaders of government and industry who have failed to respond to the challenges of keeping the United States number one in these areas.

Solutions

So what is to be done if America wants to be number one in the world in math and science? Additionally, how are we to produce more scientists and engineers—a question that may or may not be related to the United States being number one?

First, we might imitate the Japanese. And I don't mean imitating their educational system. That could prove disastrous. We should imitate how their government and business leaders have been able to marshall resources, invest in key industries, make a good product, and plan for the future. It may be that this change will have to come from the top levels of government and business management. The shortsightedness of the past three decades will have to be overcome if we are to be truly number one in scientific and technological research and development plus the production, marketing, and distribution of good quality products produced from our research efforts.

We must resist, however, imitating their educational system. As James Fallows has argued in numerous articles and in his book *More Like Us: Making America Great Again*,[16] we must capitalize on American talents and not try to artificially mimic those of the Japanese. Fallows notes that for the Japanese entrance to a prestigious university is the passport to a good life. Once in the university, students are almost assured of graduation and a good job. Thus, the pressure to enter the university is intense. Japanese students have to undergo a major series of exams at the end of their high school career, exams that place a strong emphasis on facts, facts, and more facts. As Fallows says, the most common student slogan at this point is "Pass With Four, Fail with Five!" referring to the number of hours of sleep one gets in the weeks before the exams. The system rewards memorization and punishes creativity, a fact that even the Japanese have been concerned about in the last few years. Fallows quotes the literary critic Shuichi Kato: "It's all so well designed to produce very good mediocre people. It's geared to a high level of mediocrity. The best second-rate engineers. The best second-rate workers."[17]

It's a curious dichotomy: we look at the Japanese and see their strengths, they look at us and see ours. Chalmers Johnson, in an article comparing the two educational systems, sums it up nicely:

> Japan does an excellent job with compulsory basic education, but its universities are anything but world class. In fact, many Japanese professors regard them as mediocre, particularly at the graduate level. . . . Japan educates its young children to some of the highest levels found on earth today, but it then stumbles at the post-secondary level; the United States does an inadequate job with its young children, but then it catches up with and often surpasses the rest of the world at the university level.[18]

As Fallows argues, we need to be more like us. If we are to reform the educational system, let us proceed from our strengths, from what has worked in this country. Let us look at how we have produced creative people in various areas and how our higher education system has become the best in the world. Perhaps then we could model elementary and secondary education on their strengths. Our individualism is our strength, as Fallows notes, and all methods that allow individuals to develop and learn, at any point in their lives, will produce a better workforce. We must be careful not to shut off individuality too early in life, particularly by forcing people into categories through our system of testing.

The key, then, to producing more scientists and engineers— if this is what we truly want—is to encourage more people to become interested in science and math. They don't have to take math and science courses but they should be interested in the intrinsic nature of problem solving and creativity. How are we to promote such interest?

First, I think it would be very productive to lower math requirements, especially in high school. Give students more of a chance to take their time in math courses, to have fun learning it. Give teachers more of a chance to show students practical uses of math; don't bind them so much to strict curriculum guidelines. But you say that we need increased requirements, that this is the only way to produce talented mathematics students. I believe this is a mistake; note the admonition provided by the report *Everybody Counts*, talking about what they call the myth that increased requirements yield better prepared students:

> Motivation almost always works better than requirements. Often, increased requirements have an effect quite the opposite of what was intended. In Wisconsin, for example, when the university increased from two to three years the number of courses required for admission and also increased the minimum grade point requirement, in some schools the number of students who elected four years of high school mathematics dropped. Once the three-year requirement was met, students skipped

senior mathematics to protect their grade point averages. In Florida, increased requirements for graduation from high school have caused an increase in the number who drop out.[19]

At the high school level, then, more attention needs to be placed on motivation and interest, and not on increasing requirements.

If requirements are lowered, then we must shy away from international competitions as they are presently constructed. If we take IAEP seriously, and plan to beat the other countries in math the next time around, then our thirteen-year-olds are going to have to know a lot more about this testing business. This means forcing more algebra and geometry on them when they're ten, eleven, and twelve; this would force these subjects to retreat into the middle school arenas. Not only would a major restructuring of teaching and curriculum have to take place for this to happen, but students would be faced with more math and more math pressure, precisely the conditions to produce a whole generation hostile to anything remotely mathlike. This could, in fact, decrease the number of students interested in math and science careers.

If we really want to do international comparisons for their own sake, i.e., to see how "smart" or "dumb" we are compared to other countries, and not tie it to economics, then take a fair sample, a sample of the population where it matters how smart or dumb they are: adults. Let's measure the math and science knowledge of adults who hold down jobs, who work in business or politics, or even who are unemployed. Take a representative sample of the adult population of each country. In this comparison, I'm not so sure that the United States wouldn't win, given the high level of general education in this country that has existed since World War II and is probably unequaled by many other countries in the world.

Second, if we make math in schools more interesting, we may increase the number of students who, once they enter college, are interested in math and science, or at least haven't been put off by it. Now, it is extremely crucial that colleges and universities realize the importance that freshman courses have in sustaining interest in someone's intended major. For those with a math background not strong enough to start calculus, but who are interested in science or engineering, then precalculus courses that are stimulating and not demeaning or remedial must be developed. It is the burden of the university to sustain interest. For those who do start calculus, then these courses must be changed to develop mathematical skills that are relevant to someone's intended major. Once again, the report *Everybody Counts* offers some positive suggestions for changing undergraduate mathematics:

> 1. Freshman and sophomore courses (especially calculus and linear algebra) should be taught by the most able instructors who can motivate students to study mathematics. 2. Introductory courses must be taught in a manner that reflects the era in which we live, making full use of computers as an integral tool for instruction and for mathematics. 3. Upper-division offerings should be designed to represent broadly the several mathematical sciences, introducing students in appropriate ways to applications, computing, modeling, and modern topics. 4. Diverse opportunities outside formal classroom work should be provided so that students can engage in mathematics through projects, research, teaching, problem-solving, or independent study.[20]

What would be the consequences of such changes?

> Broad undergraduate mathematics programs will attract more students to extended study of mathematics, will offer these students appealing opportunities to explore mathematics applicable

to many fields, and will engage good students in exploring and learning mathematics on their own. Such an approach, while intended as a foundation for students in many different majors, will inevitably attract good students to careers in mathematics.[21]

Third, courses and careers in math and science need to become more interesting. They may never compete with the allure of entertainment or high finance, but there are many rewarding aspects to careers that pursue the development of scientific and technical expertise. Let courses connect with these rewarding aspects. Design courses so that students engage in real activities, real problem-solving, and are able to create for themselves. This could be done with appropriate facilities at school or with schools in combination with business and research facilities. Encourage more contact between students (particularly students in elementary and high schools) and people who actually do science and math for a living. Instead of studying for so many tests, let students visit laboratories or research centers to talk with people who use math and science in their daily work. Instead of mathematicians, scientists, and engineers telling everybody that they should major in these areas, let them get off their duffs and prove why: Why should anyone spend their adult life pursuing these activities?

Finally, I would offer another solution to enticing more people into science and math careers, an alternative that I believe has been overlooked. Most of the present focus on math and science achievement is centered on high school students: make them smarter, the reasoning goes, and more will choose technical careers. The arguments above have already outlined some of the pitfalls in this line of reasoning. There is one age group, however, that has been consistently overlooked as a potential target: this is the young adult population, those roughly twenty-five to forty-five, who seem to be crowding into our nation's colleges and universities in the hope of either receiving more training for advancement in their jobs, searching for possible avenues for a change of jobs, or simply wanting an education to enhance their own lives. I have taught evening school courses to this population for over fifteen years; this group is highly motivated, studious, and serious about the task they are tackling. They know their chances are more limited, and as a consequence, they are more serious now about school than they were when they were in their teens or early twenties. They want to learn and to change their lives.

Why aren't these people attracted to science and math careers? There are probably a number of reasons. In the first place, they may be attracted to the same types of careers—e.g., business and communications—that attract the younger age groups. In many ways these are the highly attractive careers in our society. On the other hand, certain courses and majors may not be available to those in evening schools or adult programs: given the shortage of science and math teachers anyway, universities and junior colleges may be hard pressed to obtain more qualified teachers for this age group.

However, I believe there is a more profound reason that this young adult age group is not attracted to math or science: we have made such a big deal about our lack of math and science ability in this country that everybody feels "dumb." Far too many people believe they just can't do math and science. Combine this with the fact that these people will have been out of school for anywhere from five to twenty years and a fear of math and maybe science could easily have developed. If these people remember some of the math courses they had growing up, then this hostility may increase as they think about taking a math course for the first time in a long time.

We cannot afford to overlook this important age group. They are mature and incredibly energetic students and I believe they could help solve the shortage of mathematicians and scientists very quickly if we could reach them. How are we to attract them into science and math careers?

First, a major refocusing of resources would have to be undertaken. Colleges and universities would have to see the potential and get help from businesses to develop this potential. Second, the first few courses that these people take would have to be carefully structured and taught by the school's best instructors. First-semester calculus cannot be presented to someone who's been out of school for many years. The first math course would have to be entertaining, interesting, and practical: cut the math down to the essentials that these matured students are going to need to succeed in a technical career. Then have the best teachers instruct this course. The problem that I foresee here, and the challenge that I throw out, is that I don't think math departments can be so humble. Will they realize that not everyone needs two years of theoretical math to become a good scientist or technician? Will they tie their material to some practical applications and interests of their students? Will they be able to realize the psychological blocks that many people have toward mathematics and then be able to overcome them? I think that any university that can successfully answer these questions will become a model for the development of scientific talent in the 1990s.

Conclusions

I have tried to redefine what it would mean for the United States to be number one in the world in math and science. I have tried to show that being number one should have nothing to do with the way it is presently defined, via international comparisons on tests of math ability given to our thirteen-year-olds. It is not important to be number one on the test: in one sense we're really not all that far behind on school math. And to make us number one on the test would mean forcing more abstract math down the throats of young children, just for the benefit of adults who don't want to face the real problems.

We want to be number one where it counts: in business, in technological developments, and in creativity. To do so means a major shift on the part of government and business to focus on real issues of research and development in technical areas. This means a planned focus, a huge expenditure of money, and a shift to longer-range goals than we have presently considered. If we want to be number one in computers, in high definition television, or other areas, then let's go after it. But to proceed there needs to be a major refocusing at top levels of control rather than at the bottom. Finally, if we really want more scientists, mathematicians, and engineers, then let's study how motivations are developed and career choices undertaken. How does someone develop an interest in a specific area? How much is it a function of the interests of the society, how much the influence of parents or teachers, how much the development of purely personal interests? Does it have anything to do with school? Whatever the reasons, the more opportunities a society allows, and the less that society tries to pigeon-hole someone's abilities, the more possibility is allowed for people to choose any career at any point in their lives. Our universities, to some extent, have allowed this flexibility of choice. Perhaps they should serve as the model for changing America's public schools.

In thinking about these problems, we must avoid being too simplistic and in particular laying the blame for all the problems at education's doorstep. Lawrence Cremins, in a recent book *Popular Education and Its Discontents,* provides the best last word:

What about the more recent effort to use education as an instrument to achieve economic competitiveness, particularly with Japan but also with other booming capitalist economies of eastern and southern Asia? . . . American economic competitiveness with Japan and other nations is to a considerable degree a function of monetary, trade, and industrial policy, and of decisions made by the President and Congress, the Federal Reserve Board, and the federal Departments of the Treasury and Commerce and Labor. Therefore, to contend that problems of international competitiveness can be solved by educational reform, especially educational reform defined solely as school reform, is not merely utopian and millennialist, it is at best foolish and at worst a crass effort to direct attention away from those truly responsible for doing something about competitiveness and to lay the burden instead on the schools. . . The pattern bespeaks a crisis mentality inseparable from the millennial expectations Americans have held of their schools.[22]

Notes

1. George Bush, State of the Union Address, 1990.

2. Archie E. Lapointe, Nancy A. Mead, and Gary W. Phillips, *A World of Differences: An International Assessment of Mathematics and Science* (Princeton, N.J.: Educational Testing Service, 1989).

3. Ibid., p. 16. Answers: 13, 4, 7, 3, and 17.5.

4. Ibid., pp. 17–18.

5. Ibid., p. 18.

6. Ibid., p. 19.

7. William K. LeBold and Kathryn W. Linden, *Report on National Engineering Career Development Study. Engineers' Profiles of the Eighties* (Washington, D.C.: National Science Foundation, 1983).

8. National Research Council, *Everybody Counts: A Report to the Nation on the Future of Mathematics Education* (Washington, D.C.:

National Academy Press, 1989).

    9. Ibid., p. 7.

   10. Ibid., p. 39.

   11. Ibid., p. 52.

   12. Ibid.

   13. U.S. Department of Education, National Center for Education Statistics, *Digest of Education Statistics*, annual.

   14. Quoted in *Fortune* (March 26, 1990): 32.

   15. Ibid., p. 31.

   16. James N. Fallows, *More Like Us: Making America Great Again* (Boston: Houghton Mifflin, 1989).

   17. James Fallows, "Gradgrind's Heirs," *Atlantic Monthly* (March 1987): 20.

   18. Chalmers Johnson, "Economics and the Classroom: How Japan Measures Up," *Forum for Applied Research and Public Policy* 4, no. 4 (1989): 49.

   19. National Research Council, *Everybody Counts*, p. 75.

   20. Ibid., p. 53.

   21. Ibid., pp. 53–54.

   22. Lawrence A. Cremin, *Popular Education and Its Discontents* (New York: Harper and Row, 1990), p. 22.

6

Let No One Enter
Who Does Not Know Geometry

In the mid fourth century B.C., Plato, then an elderly academic at the University of Athens, was hired as a consultant to the king of Asia Minor. The king, who had read Plato's latest best seller *The Republic,* was inspired by the author's concept of the philosopher-king, an idea which appealed to him as an avenue to earthly and perhaps even immortal success. Since the king already had many regal virtues, as a consequence of his position, he needed only to acquire the tools of logic, particularly mathematics, upon which Plato had based his ideal government. Thus, Plato was dispatched to the shores of Asia Minor, with a suitable *per diem,* to teach the king and his sons the elements of geometry. After two days of drawing trapezoids, parallelograms, and squares in the sand, learning the ins and outs of interior and exterior angles, and attempting to prove theorem after theorem, the king became so frustrated that he sent Plato packing. Vowing never to waste time on math again, the king went on to conquer the rest of Asia and certain parts of Greece.

Although this first recorded example of "math anxiety" may have no bearing on Western civilization, it does illustrate how frustrating mathematics can be to learn. On the one hand are the mathematicians, the math educators, and other groups who insist that the more mathematics you learn the better off you are. Above the entrance to Plato's school was the dictum "Let no one enter here who does not know Geometry"[1]; some twenty-five centuries later, a person would be denied entrance to any major or minor college or university without a suitable background in mathematics. On the other hand, however, are the people who have struggled through years of algebra, geometry, and other types of mathematics courses with at least two questions in mind: Why I am learning this material? When will I use this stuff in real life?

Temporarily, I am going to reverse my cynicism about the value of mathematics. I am going to assume that mathematicians and math educators are right and that most people do need lots of mathematics to solve problems in everyday life. My question then becomes: Given that mathematics is important, what is the best way to teach this material and to ensure that students can apply the concepts they learn in class in situations outside of class. Through a review of the research on the teaching of mathematics in this century, I have come to some interesting conclusions. First, there is remarkable disagreement within the educational community over how to teach mathematics. These arguments revolve around several issues: Is it best to teach pure mathematics or practical mathematics? Should concepts be presented in a strict, logical progression by an expert teacher or should students be allowed to explore and discover concepts on their own? Should students be drilled on concepts or given the chance to develop their own insights into mathematical material? Should study materials be developed by teachers and other experts or can students help determine the nature of mathematics, based on their own intuitive understandings?

Second, there is little understanding of how the mathematics that is learned in school can be applied to problems faced outside the classroom. The hidden assumption behind the teaching of most mathematics is as follows: learn the basic abstract concepts, practice these concepts in the form of word problems, and then transfer your understanding to real situations. There is very little evidence, however, that this progression actually works. What, then, is the best way to learn mathematics such that it helps us solve realistic problems?

Drill vs. Insight: Thorndike vs. Wertheimer

For nearly as long as mathematics has been considered an important discipline for people to learn, there have been arguments over the best way to teach it. Almost seventy years ago, Edward L. Thorndike, one of this century's leading learning theorists and educational psychologists, criticized the techniques behind the teaching of the "old math," which had been so popular in his childhood:

> The faith in indiscriminate reasoning and drill was one aspect of the faith in general mental discipline, the value of mathematical thought for thought's sake and computation for computation's sake being itself so great that what you thought about and what you computed with were relatively unimportant.[2]

Thorndike was criticizing the teaching of arithmetic and algebra, methods that had existed in the educational curriculum of this country from the early 1880s. This approach emphasized the value of mathematical reasoning in its own right, as a subject whose rudiments could easily transfer to the learning of other disciplines; thus, algebra textbooks, for instance, dealt with "bogus" and "fantastic problems" because the content was not nearly as important as the general principles behind the content.

Thorndike did not believe in the transfer of learning, which formed the basis of the early mathematical curricula; instead, he argued for a "functional" approach. In considering the topic of algebra, for instance, Thorndike proclaimed that "algebra is a useful subject, but its utility varies enormously." To prove this, he examined all the uses of algebra in encyclopedia articles, textbooks, and scholarly articles in various disciplines. He concluded that only a small percentage of the varied techniques taught in high school classes were actually used in practical applications.

To teach algebra successfully, then, became a matter of emphasizing the practicality of the material taught. Instead of designing textbooks around the logical ordering of algebraic topics, Thorndike suggested the following:

> From the functional point of view, emphasizing ability to use algebra in solving problems which life will offer, it seems desirable to consider the lives of boys and girls and men and women as students, citizens, fathers and mothers, lawyers, doctors, business men or nurses, and select problems which they may usefully solve and which are properly solved by algebraic methods.[3]

This functionalistic philosophy had a number of implications for the construction of algebraic problems and texts. First, realistic problems should be used as much as possible. As Thorndike suggested, "Solving problems in school is for the sake of problem solving in life."[4] Second, problems should be dealt with in all their complexity and not reduced to some simple, unrealistic format, such as the way word problems often operate. Finally, students should "feel some need for a procedure and some purpose in learning it before they learn it."[5] Sections of the textbook should be divided according to such disciplines as business, social science, and science, instead of by algebraic techniques; the principles and problems to be solved may have to be relearned in each particular area, for the transfer of technique may not hold across different areas.

Not only should problems be important but they should be designed to teach what is fundamental and important to algebra. Thorndike devised lists of phenomena essential to algebra, through examinations of textbooks and talks with teachers. Instead of endless chapters of unrelated exercises, Thorndike felt that algebra could be grouped into a few basic principles, including the notion of symbolism; ability to read formulas; ability to evaluate and solve formulas, first with one unknown variable and then in two or more; and, finally, the ability to read graphs (statistical knowledge). Things to be avoided included elaborate computations, unreal or useless problems, unnecessary vocabulary, undesirable terms and definitions, superfluous connections, and shortcuts.

To teach each of these various techniques, and to insure their strength and use after the course ended, Thorndike emphasized the formation of "bonds," i.e., the learning of specific rules and places to apply these rules. Teachers should set up a particular formula, show its usefulness, establish any symbolism, variables, or abstractions involved in it, and then let the student work problems. Drill was very important for learning and should be allotted time depending on the importance of each topic. Important principles should have more time devoted to them; earlier texts had allotted equal time to all techniques, a fact that Thorndike found appalling. Spaced drill should be included: once a principle is introduced, it should be practiced at various intervals throughout the course. Thorndike felt that the reinforcing bonds would create habits that allow a student to understand when to use algebra:

> Learning to compute algebraically is not only, or chiefly, learning rules and how to apply them; it is also building up a hierarchy of habits or connections or bonds which clarify, reinforce and, in part, create the understanding of what the rules mean and when to apply them.[6]

Is learning algebra strictly the formation of habits? Are there not general mathematical principles that a student should understand that go beyond rote learning? "It so often happens that the really effective principle is the product of habits, not their producer. . . . Principles are not, as a rule, general but rather are limited to the fields where they have had habitual operation."[7] The learning of fixed rules and bonds is so important that the teaching of shortcuts can often be detrimental: "Every such rule which a pupil finds that he can break (as he can all such) without getting wrong answers means a risk that he will lose respect for and confidence in the really imperative rules."[8]

To teach effectively the entire discipline of algebra, an instructor must build all the necessary bonds, step by step, and provide enough appropriate practice, to insure the student's efficient use of techniques. Thorndike presented evidence, that, at the end of a year of algebraic drill, easy bonds, such as $3x + 5x = 8x$ could be done 99 times out of 100 in two and a half seconds or less. Achievement was emphasized, along with the fact that good bonds could produce a quick response from students. Thorndike was insistent that each bond to be acquired should be thoroughly practiced, and such practice should include problems in which the form of the bond is changed. For example, he argued, "Any disturbance whatsoever in the concrete particulars reasoned about will interfere somewhat with the reasoning, making it less correct or slower or both." To show this experimentally, he instructed students on forming the square of $(x + y)$ and the factors of $(x^2 - y^2)$, i.e., $(x + y)(x - y)$. Then he presented them with the problem of squaring $(b_1 + b_2)$ and factoring $(1/x^2 - 1/y^2)$. Although, in principle, the latter two are exactly the same as the former two, Thorndike reported that, of ninety-seven graduate students, error rates increased 6 percent to 41 percent. The "mind is ruled by habit throughout,"[9] and to insure transfer of learning one might have to teach each particular bond separately.

Thorndike believed that students should learn practical mathematics. Since the transfer of learning from one situation to another appeared difficult, he also believed that students should be constantly drilled on algebraic problems within each field, such as the algebra used in business (for instance, in calculating compound interest). Thorndike's techniques for learning mathematics were criticized for their simplistic reliance on bond formation and for not emphasizing the learning of general principles or even the meanings behind the bonds. Although he emphasized the learning of practical mathematics, Thorndike saw the student as extremely passive: He or she needs step by step help in the acquisition of material.

Max Wertheimer felt that Thorndike's students were not really learning mathematics! Representing the Gestalt school of psychology, which was to present a formidable challenge to Thorndike's functionalism and other American psychologies of the time, by challenging the notion that learning occurred strictly by the formation of bonds, Wertheimer lamented the lack of transfer of training: "What kind of education had they received, that such slight changes, which in fact did not at all touch the character of the problem, should cause so much trouble?"[10] He felt that students in Thorndike's classes were learning by rote, and did not understand the general principles behind the procedures they were learning. In his classic text on *Productive Thinking,* he referred to two types of responses, A and B, which guided reasoning. A student who evidenced B type response would learn the factors of $(x^2 - y^2)$ by rote and then be troubled by the changed conditions of $(1/x^2 - 1/y^2)$; a student who showed an A response would immediately see that the latter is only a slight variation of the former. As Wertheimer suggests, "we must differentiate, and very strongly so, between blind associations, blind habits, blind experiences, on the one hand, and the actuality of thinking, of grasping the inner relation between operations and their reasonable results."[11] Someone who has learned the principle behind factoring should be able to tell immediately that an equation like $(a^2 + 2ab + b^2)$ can be factored easily into $(a + b)(a + b)$ while one like $(a^2 + 2ab - b^2)$ cannot be factored, since there are no two numbers when multiplied that give you (-1) and when added give you $(+2)$.

In Wertheimer's scheme, the most effective learning comes from an understanding of the inner structure of a mathematics problem, or by "insight" into its fundamental nature. This can best be accomplished by letting the student "do the task, to let him find the necessary steps, but in a reasonable way, by proceeding from structurally easy tasks to structurally more difficult ones."[12] Wertheimer explored this in a number of investigations from studies of children solving simple parallelogram problems to famous mathematicians and scientists, such as Carl Friedrich Gauss (who first studied the normal distribution) and

Albert Einstein, in their discoveries of new theories.

Wertheimer pointed out two traditional types of learning that he felt were more limited than his own productive thinking: traditional logic and classical associationism. "Some psychologists would hold that a person is able to think, is intelligent, when he can carry out the operations of traditional logic correctly and easily."[13] The operations Wertheimer had in mind included the classical processes associated with deductive and inductive reasoning, in short, the processes of the logician or the geometer (cf. Plato or Euclid). Contrasted with this is the associationist approach, where thinking is the product of the contiguous succession of ideas ruled by habit, past experience, or chance such as first argued by David Hume. In opposition to both theories, Wertheimer advanced a theory of productive thinking, in which the learner is viewed not as a logic machine nor as susceptible to the random winds of past associations, but actively seeking the structures or relationships in mathematical problems. To understand subject matter means being able to handle it, to use it in all its different transformations. This recognition of structural features and transformations leads to an understanding that is not piecemeal but holistic. For transfer of learning to occur, the learner cannot be taught effectively either deductively or habitually but only productively; that is, where he or she gains insight into the structural qualities of the domain. Wertheimer emphasized the value of algebra and geometry for their own sake, a fact that Thorndike had criticized.

This clash between bond formation/rote learning and productive thinking, between habitual drill and discovery of rules, between math as practical and math as pure, became particularly strident in the 1950s and the 1960s. On the one hand, the influence of Thorndike, B. F. Skinner, and notions from behavioral psychology led to many attempts to reform curriculum and to test classroom conditions to determine the most effective factors in the learning process. On the other hand, there were repeated counter attempts to prove that drill and behavioristic methods such as programmed learning were not the most effective ways to teach mathematics. The former reforms were concerned with the best environment for learning applied and practical mathematics; the latter efforts were concerned with the most effective way of having the student learn the inner relations of mathematics to understand the essence of pure mathematics.

The "New Math": Curriculum Reform of the '50s and '60s

The dissatisfaction with the rote learning/drill method of instruction and the emphasis on learning only "practical" mathematics continued into the 1950s, when educators and psychologists began to revise theoretical notions associated with the learning process. At the same time, a historical phenomenon occurred, which, if it did not necessarily improve the teaching of mathematics, at least spurred the diverting of millions of dollars of government research money into it.

On October 4, 1957, the Russians launched Sputnik, and suddenly Americans viewed themselves as behind in the space race and in the race for world superiority. After the disastrous launch of our own Vanguard I in December (an event which the press sarcastically labeled "Kaputnik"), government officials and citizens were even more concerned with the future of Americans in science and engineering and our ability to catch up with the Russians. Many people asked the following question: how could this have happened and what can be done to prevent it from recurring?

Economic and political events can cause us to rethink how

our educational system is structured, as we have already seen in United States' responses to Japanese economic challenges in the 1980s, discussed in the previous chapter. In the late fifties and early sixties, this new concern with the teaching of science and mathematics led to two international conferences of scientists and educators on the goals of mathematics: the first at Woods Hole, Massachusetts, in 1959 and the second at Cambridge, Massachusetts, in 1963.[14] Both conferences called for a restructuring of the mathematics curriculum as the panacea for learning problems; children should be learning organized concepts and not drill in specific and arbitrary material. As Jerome Bruner remarked in 1960, in a statement from the conference proceedings that was nearly the opposite of the Thorndikian approach,

> The curriculum of a subject should be determined by the most fundamental understanding that can be achieved of the underlying principles that give structure to that subject. Teaching specific topics or skills without making clear their context in the broader fundamental structure of a field is uneconomical in several deep senses. In the first place, such teaching makes it exceedingly difficult for the student to generalize from what he has learned to what he will encounter later. In the second place, learning that has fallen short of a grasp of general principles has little reward in terms of intellectual excitement. . . . Third, knowledge one has acquired without sufficient structure to tie it together is knowledge that is likely to be forgotten.[15]

From these conferences and other interests at the time, a revolution in the study of mathematics occurred including the introduction of concept-oriented mathematics (the New Math), programmed learning, and discovery learning. The New Math proponents focused on pure mathematics, and felt that a proper understanding of mathematical material could proceed only from conceptualizing the abstract basis of mathematics. Pure mathematics took precedence over practical mathematics. Thorndike's insistence on practical mathematics was discarded and textbooks were rewritten around a conceptual mathematics, much to the chagrin of parents and students. How students were to learn the New Math centered on two approaches: drawing on the work of Thorndike and Skinner, new drill methods, in the form of a Programmed System of Instruction (PSI), were introduced; and, drawing on Wertheimer, new insight methods, in the shape of discovery learning, were also introduced.

In a major five-year study, conducted by the Minnesota National Laboratory, of the effectiveness of the new curricula versus conventional methods for teaching mathematics, Rosenbloom and Ryan[16] showed that, on all major dependent measures of mathematics ability, experimental programs did not significantly increase or decrease student performance. In a follow-up study by Ryan,[17] based on students' attitudes, interest, and perceptions of proficiency, there was once again no major difference between old and new methods.

The effects of programmed learning, a technique not originally devised by Skinner but enlarged and put to use by him, also showed no unequivocal differences in performance between PSI and conventional curriculum. Behavioral variables, such as classroom size, type of homework, and so forth, were considered to be as important as the curriculum in the teaching and learning of mathematics. The design of the efficient classroom, in the Skinnerian mode, allowed for the control of all important reinforcement contingencies such as the size of the classroom, the number of teachers in the room, and the amount of homework assigned. The manipulation of behavioral variables, however, produced very few unequivocal results in the teaching of mathematics: from homogeneous (students of the same ability) to heterogenous classes; from small to large classes; from team teaching to single teaching; from assigning homework to not assigning homework; from the use of teaching machines and programmed learning textbooks to computers. In all cases, no single method or combination of methods produced clear statistical results in terms of their effectiveness in raising students' achievement in mathematics.

Discovery learning techniques also were not proven to be more effective than other methods. Discovery learning's essential aim was to restructure mathematics by devising curricula such that students could "discover" rules and concepts instead of being taught them explicitly or by rote, such as Thorndike had argued. Thus, for instance, a student would be presented with several examples of right triangles and the relationship between two sides, and then was expected to "discern" the Pythagorean theorem. In a major review of the effectiveness of discovery learning, Shulman and Keislar remarked that "there is no evidence that supports the proposition that having students encounter a series of examples of a generalization and then requiring them to induce the rule is superior to teaching the rule first and asking the students to apply it to a wide variety of examples."[18] Furthermore, they commented on the general strategy of classroom teaching of mathematics:

> In any event, two conclusions for classroom instruction are clear: (1) There is no useful way of posing a broad question regarding how much and what kind of guidance the teacher should apply. The question should always be formulated for a specific context including the type of subject, the maturity of the pupils, prior learning experiences, and so on. (2) Regardless of which way of talking about learning by discovery is adopted, no single

teaching method is likely to accomplish the wide range of cognitive and affective objectives discussed at the conference.[19]

Piaget and the Child's Construction of Number

Amid the dissatisfaction with behavioral approaches to learning, many American educators and psychologists discovered the work of Jean Piaget in the 1960s.[20] Conducting much of his research in his native Switzerland, Piaget wanted to understand how the child's mind is different from an adult's mind and how the transitions from child to adult occur. Piaget argued for the logico-mathematical development of thinking in the child, such that the child progresses from his or her view of reality (which is prelogical and prenumerical) to the adult conception of number and logic. This development takes place in a series of four stages (sensorimotor, preoperational, concrete operational, and formal operational), with each stage represented by a certain set of operations that the child performs on the environment. A major assumption behind the growth from sensorimotor to formal operations is the child's gradual working on or restructuring to the environment or the environment to the child (the processes of accommodation and assimilation). The child cannot be shown, in most cases, the path that must be taken; instead,

In the area of logico-mathematical structures, children have real understanding only of that which they invent themselves, and each time that we try to teach them too quickly, we keep them from reinventing it themselves. Thus, there is no good reason to try to accelerate this development too much; the time that seems to be wasted in personal investigation is really gained in the construction of methods.[21]

For instance, at the sensorimotor stage, a child does not yet understand that objects have a certain permanence to them. When the mother leaves the room, for example, a child at this stage believes the mother has disappeared. At the preoperational stage, children have developed object permanence but feel that any changes to an object modifies its identiy. For example, a father with a Halloween mask on is recognized as a different, scary entity and not as the child's father. During the concrete operational stage, a child recognizes that objects can change identity and still remain the same object. Finally, at the formal operational stage, a child learns that abstract concepts can also change identity that, for instance, the concept of home can be altered (i.e., by moving), but yet still remain the same. In these examples, the child cannot be "taught" a concept but must discover the concept for himself/herself by interacting with the environment.

Before an understanding of arithmetic operations (such as addition, subtraction, multiplication, and division) can occur, certain logical antecedents must be developed first. The child must understand the notions of classification (the beginnings of concepts, words, and the properties that relate to these), seriation or ordering (that concepts can be grouped by relations, such as smallest to largest), and logical connectiveness (the operations which allow the reformulation of concepts, such as conjunction, disjunction, implicative structures, negations, and the use of symbolism, other than words). The absence of these properties is typical of the preoperational stage; in the concrete operational development the child has learned these operations and learned the idea of conservation of quantity and number, which implicitly suggests the notion of reversibility of operations. For example, take the operation of *larger than,* which is a prerequisite for addition. If a child is presented with a picture displaying a lot of poppies and a small number of bluebells, and asked which is more, he replies, "Poppies." The child is then asked if a poppy is a flower, and if a bluebell is a flower, to which he or she replies yes to both. The child is then asked which is more— the bunch of flowers or the bunch of poppies? A preoperational child will reply that the bunch of poppies is more, indicating a lack of the overall concept of flowers and the relation *larger than*. Until this is acquired, in its logical and symbolic form, the child will never learn the fundamentals behind arithmetic. Using notation, the previous example becomes P(oppies) + B(luebells) = F(lowers), which means $P = F - B$, which implies $P < F$ and $B < F$.[22]

These logical notions must develop before a child will be able to understand that the concepts of addition, subtraction, multiplication, and division are all logically related to one another, build on each other, and are logically reversible upon each other: subtraction is the reverse of addition, multiplication is simplified addition, and division is reversed multiplication. Before a child develops these concepts, all attempts at teaching will be mere exercises in verbal learning not mathematical thinking.

Two major suppositions undergird Piaget's research. First, children must invent logical and mathematical operations for themselves before such operations are truly understood. Second, most children develop these logical concepts on their own and come to a logico-mathematical view of the world by the time they reach the period of the formal operational thinking in adolescence. These assumptions have several implications for teaching and research. Teachers must not try to teach concepts before a child is ready for them. This implies an understanding of development and a willingness to structure the classroom environment such that materials are available to help the child discover and invent operations. Research into the child's understanding of mathematics can be served best by studying the development of operations and perhaps by the errors that are made. Finally, since children invent these operations, it becomes fruitless to try and train them in these concepts or try to speed up their development. As Piaget remarked:

> This question (of accelerating learning) never fails to amuse students and faculty in Geneva, for they regard it as typically American. Tell an American that a child develops certain ways of thinking at seven, and he immediately sets about to try to develop those same ways of thinking at six or even five years of age. Investigators in countries other than America have tried to accelerate the development of logical thinking, and we have available today a considerable body of research on what works and what doesn't work. Most of the research has not worked. It hasn't worked because experimenters have not paid attention to equilibrium theory. The researchers have tried to teach an answer, a particular response, rather than to develop operations. They have tried to teach the child that, of course, the hot dog (shaped piece of clay) will weigh as much as the clay bell; just put both on a two-pan balance and you'll see. But the child is completely unconvinced unless he shuffles the data around in his mind, using one or more of the operations I've described. Learning a fact by reinforcement does not in and of itself result in mental adaptation.[23]

The "New Mind": The Cognitive Revolution

Although Gestalt psychologists had emphasized mental qualities such as insight, American concerns with "mind" and "consciousness" came much later. The cognitive revolution of the sixties and seventies, spurred on not only by an interest in Piaget's work but also by the use of computers as a model for human thinking and reasoning, explored and reassessed many topics in the learning of mathematics. Mathematics provided a very appropriate model for use on the computer; the logic, discreetness, and well-formed nature of mathematical problems provided the tools for simulation and testing of theories about the learning of mathematics. Furthermore, the reappraisal would be influenced by critiques of past research; Thorndike had emphasized the formation of bonds but had done little research on which bonds were easy to acquire, which were difficult, and why; Wertheimer had explored errors that people make in certain types of geometry problems, but his investigations had been limited; and the decade of curriculum reform in the sixties had emphasized a restructuring of the content of mathematics, but the assessment of effectiveness had always been in terms of global scores on "math" tests, with little concern for specific problems in specific areas for specific learners.

Cognitive researchers sought to remedy the lack of knowledge about mathematics *per se* and individual learners; in Resnick and Ford's words, "It is this dual knowledge—knowledge of the structure of mathematics and knowledge of how people think, reason, and use their intellectual capabilities—that furnishes the ingredients for a psychology of mathematics."[24] In studying mathematics, researchers would now attempt to offer theoretical models that could account for three factors: (1) correspondence— how much one's mental picture of a mathematical phenomenon mapped onto the correct mathematical concept; (2) integration— how much interrelatedness there was between different mathematical concepts; and (3) connectedness—how much knowledge of one domain transferred to knowledge of other domains. Most of the research in cognitive psychology and cognitive science in the past fifteen years has addressed these concerns.

For example, Suppes and Groen[25] studied elementary school

children's ability to solve simple addition problems of the form $2 + 7 = ?$. The authors postulated three models by which this problem could be solved. Model A stated that the addition would be carried out by starting at 0; add 2; then add 7 and report result. Model B stated that you start at 2; add 7; and report result. Model C claimed that a person started with the larger unit (in this case 7); and then added the smaller (or 2) and report the result. In studies using all possible combinations of two digits, the authors found that Model C provided the more accurate match with data. When confronted with a simple addition problem, students tend to start with the larger number and then add the smaller to it.

Lankford[26] studied children's solution processes in addition, subtraction, multiplication, and division. He claimed that the errors of most pupils are highly individualized but often systematic, i.e., they follow a logic unique to the student. For example, in multiplying 304 by 6 the answer written down was 1804, where the student failed to carry the 2. From talking to this student, however, a different logic emerged: he had multiplied 6 by 4 equals 24; written 4; carried 2; multiplied 0 by 2; written 0; and then 6 by 3 equals 18—for an answer of 1804. Here the problem was not a technical one in carrying, but one in knowing what to do with what is carried.

The fact that addition and subtraction were not seen as complementary processes was demonstrated by Greeno and Resnick.[27] Greeno argued that, in theory, addition and subtraction are parallel processes; the same mental process should see that $m + n = mn$ can be reversed to produce $mn - n = m$. Students, however, only gradually understand these connections. In Resnick's example, a student is given the subtraction problems of $36 - 27 = ?$ and $27 - 36 = ?$ In both cases, one child's algorithm consisted of subtracting the large number from the smaller one in each column (producing an answer of 11 for both problems), indicating a misunderstanding of the process of subtraction.

Although research of this type has been extended to other mathematical domains, its primary emphasis has been on simple phenomena and the representation of solution strategies within these simple phenomena. Errors are seen as systematic, if idiosyncratic; the lack of interrelatedness of concepts is seen as a lack of logical connectiveness between concepts, a factor that may be susceptible to training. Currently, the cognitive science revolution is seen as the primary mover for research in the learning of mathematics, with its emphasis on the subject matter of mathematics, applied concerns, and individual learning processes.

The applications and evaluations of this cognitive approach have yet to be tested. The emphasis on the "new mind" has, to this point, been with small concerns—simple problems of addition, subtraction, geometry, etc. The large-scale factors of motivation in learning and the issue of transfer of training have only begun to be addressed. As Romberg and Carpenter comment:

> There have been attempts to draw instructional implications from recent research in cognitive science but most of the implications are still in the potential stage, and much of the research directly addressing questions of instruction has remained untouched by the revolution in cognitive science.[28]

Seymour Papert and the Children's Machine

"Do you know how giraffes sleep?" Jennifer, a four-year-old pre-schooler, had asked Seymour Papert this question because she heard that he had grown up in Africa. Papert admitted that he didn't know the answer; he asked Jennifer and her friends if they knew. They proceeded to offer several theories: "Perhaps a giraffe sleeps like a horse, standing up"; "Maybe they cuddle up like dogs"; or "It finds a tree with a branch to rest its neck." When Papert returned home, he turned his library upside down, trying to find the answer to this question.[29]

Papert suggests that children love to construct theories about the world. At an early age, they seek to make sense of the world around them and, in so doing, often pose questions that challenge adults' knowledge. Drawing on the cognitive science research discussed in the previous section, Papert notes that children are always attempting to construct their own versions of the world; in the process they often explore and develop much more complex renditions of reality than they encounter in school.

What has been lacking in the past are ways for children to test their theories easily. In the past, they have had to rely on the adults around them: "In this conversation we see two sides of the intellectual life of children of this age; the coexistence of a remarkable capacity for making theories with a nearly helpless dependence on adults for information that will test the theories or otherwise bring them into contact with reality."[30] With the introduction of computers, electronic communications, and the coming worldwide information superhighway, Papert suggests that children will no longer have to rely totally on adults. They will have access to what he calls the "Knowledge Machine"— the ability to search almost unlimited amounts of information through technical controls that exist in the home.

This Knowledge Machine should change the nature of school and how students learn mathematics. Papert contrasts two visions of the educational process—Schoolers versus Yearners. Schoolers insist that the present crisis in education will be remedied if schools are "fixed"—standards tightened, students required to do more homework, a longer school year, and so forth. To Schoolers, knowledge is transmitted as follows:

Knowledge is made of atomic pieces called facts and concepts and skills. A good citizen needs to possess 40,000 of these atoms. Children can acquire 20 atoms per day. A little calculation shows that 180 days a year for 12 years will be sufficient to get 43,200 atoms into their heads. . . . It follows that the technicians in charge (hereafter called teachers) have to follow a careful plan (hereafter called the curriculum) that is coordinated over the entire 12 years. . . . Teachers can be supervised by curriculum coordinators and department heads, these by principals, and these in turn by superintendents.[31]

Yearners, on the other hand, feel that it is possible for something totally different from school to exist. They feel that ways of knowing are more related to asking questions than memorizing fixed curricula. They feel that experimentation and multiple answers are more important than a correct answer to a test question. In his earlier research, Papert was drawn to why children play videogames so intensely and sometimes ignore their homework. As he suggests, video games are more for Yearners than for Schoolers:

Video games teach children what computers are beginning to teach adults—that some forms of learning are fast-paced, immensely compelling, and rewarding. The fact that they are enormously demanding of one's time and require new ways of thinking remains a small price to pay (and is perhaps even an advantage) to be vaulted into the future. Not surprisingly, by comparison school strikes many young people as slow, boring, and frankly out of touch.[32]

In terms of the teaching of mathematics, Papert is strongly opposed to formalized systems of instruction that insist on children learning a set number of mathematical concepts. As

Papert notes: "It is simply double-talk to ask children to take charge of their own learning and at the same time order them to 'discover' something that can have no role in helping them understand anything they care about or are interested in or curious about."[33] School ways of teaching mathematics even limit the creativity of teachers: "As long as there is a fixed curriculum, a teacher has no need to become involved in the question of what is and what is not mathematics."[34]

Papert has developed a computer language called Logo, which allows children to become software designers and in the process develop nonformalized, intuitive knowledge of mathematics. His language allows children to explore mathematical questions and develop knowledge that is connected with their own intuitive understandings of the world. For instance, before he let them use Logo, Papert asked students "What is a fraction?" Some answered that they hadn't done that yet in class. Or when asked for examples, one student's replies were limited to textbook examples, such as a physical piece of a physical thing, like a piece of pie.

After using Logo, this particular student's theory of fractions expanded to more everyday knowledge. Anything that you can think of can be an example of a fraction, she said: half an hour, twenty-five cents, a half-price sale, daytime. As Papert notes:

> The point is that formal school knowledge of fractions was not connected with her intuitive everyday knowledge. What she learned in class was brittle, formal and isolated from life. Attempts by teachers and textbook authors to connect school fractions with real life via representations as pies simply resulted in a new rigidity.[35]

Papert's philosophy is a curious mixture of positions that have been discussed in this chapter. He is certainly in the tradition of Piaget and the cognitive psychologists who argue for children inventing or constructing the knowledge of the world around them. In terms of mathematics, he is strongly opposed to mathematical knowledge being determined by school authorities, as is often represented in traditional textbooks. He is also even more practical than Thorndike was, in terms of suggesting that children need mostly to connect mathematical knowledge to their everyday world. If his Knowledge Machine comes into existence, then the whole notion of teaching mathematics in school may become an anachronism!

Math in School and in the Supermarket

Which of the following are the best buys at the supermarket?

A jar of Brand A jam that sells for $1.50 for 18 oz. or a jar of Brand B jam that sells for $1.05 for 12 oz.

A box of Brand A raisin bran that sells for $1.58 for 20 oz. or a box of Brand B raisin bran that sells for $1.13 for 15 oz.[36]

Jean Lave has tried to understand how arithmetic calculations are done in everyday settings such as the supermarket. She is also interested in whether or not what is learned in school (fractions, decimals, and procedures for solving ratios) helps someone solve a problem that they might encounter in everyday life. In the psychological literature on problem solving, this ability to apply knowledge from one situation to another has been termed "transfer of training." As Lave notes, current educational philosophy is built on the assumption that children can be taught general cognitive skills (e.g., reading, writing, mathematics, critical thinking) abstractly and "disembedded from the routine contexts of their use."[37] These general cognitive skills can then be transferred to general application in all situations. Tests of these general skills do not have to be practical in any sense but need only measure the general principles.

Lave questions whether school mathematics and everyday mathematics operate in the same way. For instance, a school math problem would be as follows: "Becca has four apples and Maritza has five apples, how many apples in all?" The answer to this problem is, of course, nine. But what does an apple problem look like at the supermarket? The following is how one shopper solved this problem:

> There's only about three or four [apples] at home, and I have four kids, so you figure at least two apiece in the next three days. These are the kind of things I have to resupply. I only have a certain amount of storage space in the refrigerator, so I can't load it up totally. . . . Now that I'm home in the summertime, this is a good snack food. And I like an apple sometimes at lunchtime when I come home."[38]

There are several possible answers to this problem, depending on which conditions are considered in the purchase of the apples.

Lave has investigated the arithmetic practices in everyday life in a project she termed the Adult Math Project. To conduct her research, Lave and her associates followed participants to the supermarket: "We followed participants at times when they fit shopping into their schedules (any hour of the day or night). We arrived at the house in time to observe preparations for shopping, went to the store together, shopped, and returned home to follow the process of storing groceries as the expedition ended in their kitchens."[39] At the supermarket, Lave watched and took notes as participants did comparison shopping. Back at their homes, Lave presented adults with a variety of math problems: from best buy supermarket problems (see the two examples above) to tests of school mathematics.

What were her results? On supermarket problems, she found that adults were extremely accurate in selecting the best buys. Although they would use a variety of strategies to solve the problem (not just the "correct" approach of school mathematics), the participants could usually arrive at the best choice. On the other hand, however, these same adults were only average at solving arithmetic and fraction tests from school mathematics. They could solve very difficult, realistic problems but had much more difficulty with abstract problems. Furthermore, the number of years the adults had spent in school related to how well they did on school math tests but not to their ability to solve the supermarket problems. In other words, no matter how much formal schooling had been completed, the adults in this study were very good at solving real problems.

Let's look at the jars of jam problem above. Shoppers could determine that the larger jar ($1.50 for 18 oz) was the best buy; for instance, reducing the amount by 1/3 to 12 oz yielded a price of $1.00 which is a better buy than the alternative jar, which is price at $1.05 for 12 oz. In effect, the shoppers have solved a complicated 3/2 ratio problem. The raisin bran problem is even more complicated. In this example, some shoppers would compute an approximate unit price; for instance, the larger box at 20 oz for $1.58 is about $.08 an ounce. If the smaller box of raisin bran at 15 oz. were priced similarly, it should cost about $1.20. But since it's priced at $1.13, it is probably the better buy. Thus, in effect, a 4/3 ratio problem has been solved by estimation.

Adults who could solve these problems, however, had a great deal of difficulty with a school fractions problem. For instance, which of the following is the larger fraction: 8/13 or 4/7? A school math solution would be as follows: calculate a common denominator for the two fractions, by multiplying the two denominators (7 × 13 = 91); convert the fractions to this common denominator (8/13 = 56/91 and 4/7 = 52/91); compare these two fractions to see which is largest (in this case, the answer is 8/13).

If the problem were rephrased as follows, would it be easier? Brand A is available for 4 oz. at $0.70, Brand B at 8 oz for $1.30. Which is the better buy (the larger fraction)? Is it clearer that Brand B is the better buy? For Brand A, a shopper would have to spend $1.40 to receive 8 oz., whereas Brand B offers the same amount for $1.30. Adults often fail the school math problem but usually succeed at the more practical exercise.

Lave's research is just one example of many current efforts that question whether general cognitive skills, such as mathematics, transfer from one situation to another. Lave questions whether there is one universal form of mathematics that can be transported to all settings and its principles used uniformly to solve any problems. As her results demonstrate, it is not easy to conclude that the mathematics learned in school will be the mathematics used in real life.

Textbooks, Tests, and Teachers

What is the "best" method to teach "mathematics"? As I have tried to show, this has been an extremely difficult question to answer. Part of the problem lies in the question "What is mathematics?" To some educators, mathematics has to be pure mathematics, grounded in the logical conceptual structures of the discipline, and taught step by step in a logical progression. To other researchers like Thorndike, the essence of mathematics rests in its practicality; there is only a need to teach the algebra or geometry that would be used in everyday life. Seymour Papert would probably want to shelve all of school mathematics, whether it is focused on pure or practical applications. To him, students must construct an intuitive sense of mathematical knowledge from their own experiences.

Other difficulties lie in searching for the "best method" of instruction. Should students be treated as passive creatures and drilled with mathematical material followed by the proper reinforcement? Or can students be more active learners, perhaps discovering mathematical concepts by reasoning from well-chosen examples? Will not the answers to all of these questions influence what kind of mathematics is offered to students at all grade levels?

In one sense these questions may matter very little, because school mathematics seems to exist in a world separated from the academic research on mathematics education. In many ways, the most important matters for teachers and students in the classroom, from kindergarten through the twelfth grade, are the assignments in the textbook and the tests that must be taken regularly. What type of mathematics is presently taught in the classroom?

In a study for the National Science Foundation, George Madaus[40] and his colleagues at Boston College examined the content and skills covered by the most commonly used textbooks and standardized tests in the United States today. They examined both mathematics and science textbooks and tests, ranging from grades four to twelve. They rated the textbooks and tests for the types of mathematics skills they presented; furthermore, they questioned teachers on the ways in which textbooks and tests influenced their teaching in the classroom.

According to Madaus and his colleagues, both mathematics textbooks and standardized tests emphasized low-level skills in mathematics, much like the abstract mathematical word problems that have been discussed throughout this book. As they note: "In math standardized tests, for example, only 3% of the questions tested high level conceptual knowledge, and only 5% of the questions tested high level thinking skills such as problem solving and reasoning."[41] The textbooks were equally limited: "95% of the items sampled recall of information, computation, and use of algorithms and formulas in routine problems similar to those in texts."[42] It is safe to assume that more creative problems or those addressing practical math, such as the problems advocated by Papert and Lave, were not to be found.

No matter what their teaching method, teachers indicated that they relied extensively on material in the textbooks in their instruction to students. Furthermore, many of the tests administered in the classroom were taken directly or derived from those presented in the textbooks. In districts where student performance on standardized tests was heavily emphasized, teachers relied even more on set curriculum guides and textbook information. More creativity in teaching mathematics was mainly evidenced in schools that deemphasized standardized tests: "In interviews we found that, in districts and schools with less emphasis on standardized test scores, and without such severe penalties attached, teachers felt more at liberty to adopt innovative curriculum materials designed to support higher-order thinking, and were doing so."[43]

If school mathematics is still emphasizing such low-level mathematics skills, over a period of eight years in grades four through twelve, then the prospects for students learning mathematics that can be applied in everyday life seems dismal. Perhaps, as Jean Lave's research demonstrates, adults will invent their own mathematics strategies when called on to solve real problems.

Or if Seymour Papert's Knowledge Machine comes to fruition, adults can assess information and expertise whenever it is needed.

At some point it seems that a choice has to be made: who is going to decide what will be learned in school and how it is to be studied? The next chapter addresses the efforts at mathematics reform in the 1980s and the 1990s and seeks an answer to this question.

Notes

1. Will Durant, *The Life of Greece* (New York: Simon and Schuster, 1966).

2. E. L. Thorndike, M. Cobb, J. S. Orleans, P. M. Symonds, E. Wald, and E. Woodyard, *The Psychology of Algebra* (New York: The Macmillan Co., 1926), p. 96.

3. Ibid., p. 109.

4. Ibid., p. 154.

5. Ibid., pp. 152–53.

6. Ibid., p. 246.

7. Ibid., p. 245.

8. Ibid., p. 239.

9. Ibid., p. 458.

10. Max Wertheimer, *Productive Thinking*, enlarged edition, ed. Michael Wertheimer (New York: Harper and Row, 1959), pp. 153–54.

11. Ibid., p. 167.

12. Ibid., p. 164.

13. Ibid., p. 6.

14. See D. J. Dessart and H. Frandson, "Research on Teaching Secondary School Mathematics," in *Second Handbook of Research on Teaching*, ed. Robert M. W. Travers (Chicago: Rand McNally Co., 1973). See also L. B. Resnick and W. W. Ford, *The Psychology of Mathematics for Instruction* (Hillsdale, N.J.: Lawrence Erlbaum, 1981).

15. Resnick and Ford, *The Psychology of Mathematics for Instruction,* p. 104.

16. P. C. Rosenbloom and J. J. Ryan, *Secondary Mathematics Evaluation Project: Review of Results* (St. Paul, Minn.: Minnesota National Laboratory, 1968).

17. J. J. Ryan, *Effects of Modern and Conventional Mathematics Curricula on Pupil Attitudes, Interests, and Perception of Proficiency* (Washington, D.C.: United States Department of Health, Education, and Welfare, Bureau of Research, 1968).

18. L. S. Shulman and E. R. Keislar, eds., *Learning by Discovery: A Critical Appraisal* (Chicago: Rand McNally Co., 1966), p.191.

19. Ibid., p. 186.

20. See Jean Piaget, *The Child's Conception of Number* (London: Routledge and Kegan Paul, 1956). See also Jean Piaget, *The Early Growth of Logic in the Child* (New York: Harper and Row, 1964).

21. Quoted in R. W. Copeland, *How Children Learn Mathematics: Teaching Implications of Piaget's Research* (New York: The Macmillan Co., 1970), p. 22.

22. Ibid., p. 143.

23. Ibid., pp. 20–21.

24. Resnick and Ford, *The Psychology of Mathematics for Instruction,* p. 4.

25. P. Suppes and G. J. Groen, "Some Counting Models for First-Grade Performance Data on Simple Addition Facts," in *Research in Mathematics Education,* ed. J. M. Scandura (Washington, D.C.: National Council of Teachers of Mathematics, 1967). See also G. J. Groen and J. M. Parkman, "A Chronometric Analysis of Simple Addition," *Psychological Review* 79, no. 4 (1972): 329–43.

26. F. G. Lankford, *Some Computational Strategies of Seventh Grade Pupils,* Final Report, Project No. 2-C-013 (Department of Health, Education, and Welfare, National Center for Educational Research and Development and the Center for Advanced Studies, University of Virginia, 1972).

27. J. G. Greeno, "Analysis of Understanding in Problem Solving," in *Developmental Models of Thinking,* ed. R. Kluwe and H. Spada (New

York: Academic Press, 1980); and J. G. Greeno, "A Study of Problem Solving," in *Advances in Instructional Psychology,* ed. R. Glaser, vol. 1 (Hillsdale, N.J.: Lawrence Erlbaum, 1978); L. B. Resnick, "The Role of Invention in the Development of Mathematical Competence," in *Developmental Models of Thinking,* ed. R. Kluwe and H. Spada (New York: Academic Press, 1980).

28. T. A. Romberg and T. P. Carpenter, "Research on Teaching and Learning Mathematics," in *Handbook of Research on Teaching,* ed. Merlin C. Wittrak (Chicago: Rand McNally Co., 1986), pp. 850–73. See also A. H. Schoenfeld, *Cognitive Science and Mathematics Education* (Hillsdale, N.J.: Lawrence Erlbaum, 1987).

29. Seymour Papert, *The Children's Machine: Rethinking School in the Age of the Computer* (New York: Basic Books, 1993).

30. Ibid., p. 7.

31. Ibid., p. 62.

32. Ibid., p. 5.

33. Ibid., p. 16.

34. Ibid., p. 79.

35. Ibid., p. 109.

36. Jean Lave, *Cognition in Practice: Mind, Mathematics, and Culture in Everyday Life* (Cambridge: Cambridge University Press, 1988), p. 104.

37. Ibid., p. 8.

38. Ibid., p. 2.

39. Ibid., p. 49.

40. G. F. Madaus, M. M. West, M. C. Harmon, R. G. Lomax, and K. A. Viator, *The Influence of Teaching Math and Science in Grades 4-12. Executive Summary* (Boston: Center for the Study of Testing, Evaluation, and Educational Policy, Boston College, 1992).

41. Ibid., p. 2.

42. Ibid., p. 12.

43. Ibid., p. 4.

7

How Much Mathematics
Is Really Needed?

In this chapter, I will argue that there should be neither national standards nor a national assessment in mathematics. Contrary to prevailing political and educational rhetoric, I believe that the institutionalization of prescribed material to be learned, followed by a single examination, will prove counterproductive. If these national efforts are not halted, students will not learn how to apply mathematics to real problems, both personal and societal. Instead, they will be instructed in material that very few will ever need again in their lives.

Furthermore, mathematics should not be taught as a separate discipline, particularly at the high school level (where separate courses in algebra, geometry, trigonometry, calculus, and statistics dominate curricula). The average citizen in our democracy will not use the abstractions of algebra, geometry, and trigonometry in his or her job. And, as I have shown, there is also no evidence that mastering analytic skills in mathematics transfers to competent logical reasoning in other domains.

If these traditions and assumptions are altered, what would the high school curriculum look like without mathematics courses? The inescapable conclusion is that mathematics instruction would be integrated with instruction in other areas. Students would practice in school what they will eventually practice in life. Invariably this means solving or attempting to solve complex problems, processes that draw on knowledge from various disciplines. As I have argued, learning how to transfer mathematical knowledge to nonschool situations is a difficult and challenging problem. Only by facing this complexity can teachers help students learn mathematics and other material that they will need in everyday life.

In the sections that follow, I will develop my case against national standards and a national examination in mathematics. I will also assess efforts to consider seriously the abilities and problem-solving skills students should be learning in order to prepare them for life in our democracy.

The Argument Against National Standards

Who would argue for national standards in mathematics, and why? In 1989 the National Council of Teachers of Mathematics issued an extensive document entitled *Curriculum and Evaluation Standards for School Mathematics*.[1] Developed over several years by representatives of the NCTM and other organizations, the standards provide detailed guidelines for what students should know about mathematics, how it should be taught, and how it should be assessed. The NCTM's vision is "to establish a broad framework to guide reform in school mathematics in the next decade."[2]

Why the need for new standards? The standards appeal to the transformation of American society in the late twentieth century from an industrial society to an information one. Schools are no longer training workers for factories and farms; students no longer must learn standard bits of accepted knowledge at minimal levels; and students no longer must learn to work quietly on assembly lines doing routine tasks day after day. The factory model of education is no longer adequate for educating workers for today's tasks. As Edward Fiske has noted:

> Public schools are nineteenth-century institutions because they were organized around an industrial model that prevailed at the turn of the century. Mass production sought to reduce as many elements of the manufacturing process as possible to simple, repetitive tasks that could be handled by workers who were easily trained and, for all practical purposes, interchangeable. A relatively small group of people—perhaps 20 percent—did the thinking for the entire enterprise.[3]

More and more, jobs require an entirely different workforce, one that is more flexible, more capable of handling different problems regularly, and one able to manipulate information more readily. Citing a study by the U.S. Congressional Office of Technology Assessment, the authors of the standards note that "employees must be prepared to understand the complexities and technologies of communication, to ask questions, to assimilate unfamiliar information, and to work cooperatively in teams. Businesses no longer seek workers with strong backs, clever hands, and 'shopkeeper' arithmetic skills."[4]

To accomplish this transformation, the authors of the standards argue that four new social goals for education must be adopted: (1) the development of mathematically literate workers who understand how to set up problems, how to use a variety of techniques to approach and work on problems, and who have the ability to work with others; (2) the promotion of lifelong learning, in which students learn to explore, create, and adapt to changing situations; (3) the provision of opportunities such that all groups have the chance to become mathematically literate; and (4) the nurturance of an informed electorate, one that can thoughtfully explore complex issues such as environmental protection, nuclear energy, and taxation. To understand these problems requires "technological knowledge and understanding. In particular, citizens must be able to read and interpret complex, and sometimes conflicting, information."[5]

It is hard to disagree with these goals. I agree that the ability to deal with complex problems is one of the most beneficial results that students can receive from an education in our society. The standards, however, explicitly connect the ability to accomplish these goals with the learning of mathematics: their assumptions are that students should have numerous mathematical experiences, develop mathematical habits of mind, value the mathematical enterprise, and "be encouraged to explore, to guess, and even to make and correct errors so that they gain confidence in their ability to solve complex problems."[6]

The belief that mathematics is the "best" of all the disciplines, the most fruitful for both intellectual and practical success in our democracy, is only thinly disguised in the rhetoric of the standards. The belief that mathematical thinking is intimately connected to the development of more logical thinking (as has been challenged throughout this book, ranging from Paulos's assertions to the claims of the SAT) is even more explicitly stated. The standards suggest that by partaking of the varied mathematical experiences they have outlined, a student will gain "mathematical power," a concept that "denotes an individual's abilities to explore, conjecture, and reason logically, as well as the ability to use a variety of mathematical methods effectively to solve nonroutine problems." Furthermore, mathematical power is "based on the recognition of mathematics as more than a collection of concepts and skills to be mastered; it includes methods of investigating and reasoning, means of communication, and notions of context." Finally, the acquisition of mathematical power will lead to the "development of personal self-confidence."[7]

The authors of the standards suggest that classrooms will change as students explore more interesting problems. Students' values, self-esteem, and ability to reason will change dramatically as they explore more challenging mathematical concepts. For instance, students will learn to value mathematics and its interaction with our culture and history; students will become confident in their own abilities, as the study of mathematics transfers to their understanding of more complex problems in the world around them; students will become mathematical problem solvers, a skill "essential if [they are] to be . . . productive citizen[s]"[8]; students will learn to communicate mathematically, through the use of a symbolic language that will help to elucidate their own thinking processes; and, finally, students will learn to reason mathematically, a skill at the heart of more general problem solving.

To the credit of the standards' authors, they also argue for a presentation and incorporation of mathematics into the classroom that is not entirely rooted in the belief that mathematics is essentially rote memorization of information. They argue that "doing" mathematics is the key to understanding and using concepts and procedures. Additionally, they note that some aspects of mathematics have changed in the last decade, particularly in the analysis of quantitative information in the social and life sciences. The ideas in these disciplines are not entirely addressed in the traditional engineering sequence of classes, consisting of algebra-geometry-precalculus-calculus. Finally, they recognize that technology is needed to do mathematics, particularly the use of calculators and computers.

What, however, will be the substance of these standards? When teachers prepare to teach and students do homework, what exactly will be the shape of these new mathematics? I see three problems with the supposed transformation of how mathematics is taught: (1) The underlying philosophical and pragmatic belief is still that mathematics is the supreme discipline, essentially invaluable for someone to survive. Without mathematics, life as a productive citizen is somehow diminished. Anyone who falls short of these "standards" is evidently not quite capable intellectually, productively, interpersonally, and may even lack self-confidence. (2) Very little, if anything, is given up by the mathematicians in advocating these standards. What was already a crowded high school curriculum has become even busier: most of traditional mathematics is retained in the standards, and newer topics are added as well. The standards' authors imply that the more mathematics students take, the better off they will be. (3) Finally, the word problem survives! The instantiation of these new standards, the actual types of problems that students must solve, has changed little. It is almost as if mathematicians, despite themselves, cannot break the frames of mind which produce problems that to them seem practical but to others seem irrelevant.

I support these assertions by reviewing the essentials of some of the standards. In the following discussion, I focus on the standards for mathematics in grades nine through twelve.

Let us examine one new problem first.[9]

Real-world problem situation. *In a two-player game, one point is awarded at each toss of a fair coin. The player*

*who first attains n points wins a pizza. Players A and
B commence play; however, the game is interrupted at
a point at which A and B have unequal scores. How should
the pizza be divided fairly? (The intuitive division, that
A should receive an amount in proportion to A's score
divided by the sum of A's score and B's score, has been
determined to be inequitable.)*

This particular problem is introduced in Standard 1, which
discusses Mathematics as Problem Solving. The goals of this
standard are to increase students' confidence in doing mathematics, particularly in applying mathematical problem solving
to situations within and outside of mathematics. In particular,
students should be able to "apply the process of mathematical
modeling to real-world problem situations." As the standard
suggests, mathematical modeling is increasingly being used in
various fields, such as the physical and social sciences, business,
and engineering; not to mention its application to "questions raised
in everyday life."[10]

I am not disputing the assertion that mathematical modeling
techniques are being widely used in certain disciplines, with
varying degrees of success. The difficulty, once again, is with
the reliance on word problems. This pizza example raises many
questions: What kind of game is this? Who would play this game?
What type of pizza are they playing for? How did they get
interrupted? Why don't they just resume the game? Most important, why is the intuitive division of the spoils not acceptable?
Who has determined that this intuitive split is inequitable and
for what reasons?

The mathematically precise and correct answer that is presented to this problem only reinforces the assumption that
mathematical problem solving is deterministic, i.e., operating by

a fixed set of rules leading to *one* correct answer. Furthermore, it underplays any resort to intuition or alternative techniques. Since I have personally been involved with mathematical modeling in the social sciences, it overlooks the innumerable difficulties (not to mention the quality) of using such techniques to solve real problems.

Here is another example, this one involving a consumer application of mathematics:[11]

> *A student deposits $100 in a savings account earning 6% interest compounded annually. How much money will be in the account at the end of 10 years?*

The standards emphasize different levels of difficulty in solving this problem mathematically. Students can simply start calculating by hand the interest and carry the arithmetic forward for a ten-year cycle. A computer spreadsheet could help simplify the process. At higher levels, students could perhaps derive the formula for compound interest, explore alternative values for years, different interest rates, and so on, and perhaps rearrange the formula to solve for any term they want.

This problem, however, has no realistic context. There are many computer programs which will calculate the answer to this problem and any of its variations, in less than a heartbeat. Hence, the given answer to this problem is trivial, at best. By exploring how the compound interest formula is derived, will students have a better understanding of realistic problem solving procedures? Even in this simple problem certain questions arise: Who is saving this money and for what reasons? Why ten years? Do interest rates really stay constant for a ten-year period? An interesting practical variation on this problem would be to visit the local bank and ask them, "If I deposited $100 in your bank ten years ago, how much would I have today?" Would this result match the formulaic answer? If not, why not?

I would offer a more realistic problem:

I own a house and I've been given offers to refinance my first and second mortgage. Several different mortgage companies have approached me for my business. Each one offers different interest rates, length of mortgages, closing costs, points, insurance, appraisal fees, insurance coverage, and certain other applicable fees. Each vendor requires different materials from me: credit histories, salary verifications, income tax returns, and a personal financial statement. Several of my friends have dealt with the different companies. Each one has a different opinion as to the ability, honesty, speed, and accuracy of the different vendors. I personally like some of the representatives I've talked to and don't like some of the others. Now, in the event that I refinance my house, what factors do I need to consider in choosing a vendor? (Maybe I should consider moving to an apartment?)

This problem is much more contextual and presents issues associated with realistic mortgage negotiations. The computation of costs, however important, is only a minor part of the decision making process. Mathematicians, however, cannot escape from themselves. Mathematical word problems continue to exist in a separate space-time continuum, completely oblivious to the realistic concerns of students. As the contrast between my problem and theirs suggests, there is a world of difference between the supposed simplicity of mathematical algorithms and the incredible complexity that confronts the average citizen on a complex issue such as buying and maintaining a house. Michael

Apple supports this criticism of the unreality of the standards: "Examples based on job loss, on the lowering of wages and benefits, on the cutbacks in welfare payments that conservative governments are forcing on already poor parents—each example and problem perhaps centered on how mathematics can help us understand what the effect of all this means for health care (or lack of it), nutrition, the family's finances, and even on the budget and resources of the very schools the students attend—these kind of problems would have been powerful ways of linking mathematics to the real world of those students who are least likely to succeed in school."[12]

A second problem I have with the standards is that the mathematical curriculum, particularly at the secondary level, has burgeoned beyond all reasonable proportions. The standards argue for enrichment and extensions of the content already covered in traditional high school courses, instead of any deletions. They argue that three years of high school mathematics should be required for all students and four years of study for "college-intending" students. This additional material is intended primarily for students who will attend college; however, since all mathematics is worthwhile, "we believe that these additional curricular topics should be studied by all students who have demonstrated interest and achievement in mathematics."[13]

What constitutes this new curriculum? Our old friends, algebra, geometry, and trigonometry are steadfastly in place, with certain topics to receive increased attention and some to receive decreased attention (but no deletions). Furthermore, students would be confronted with new sections or courses in functions, statistics, probability, discrete mathematics, and the conceptual underpinnings of calculus. Across all topics, teachers would be expected to emphasize how mathematics can be used for problem-solving applications, to show how mathematics extends logical reasoning skills, to demonstrate the interconnections of mathematics to other disciplines, and show how mathematics can be used to communicate ideas in symbolic terms.

Finally, my third problem is with the arrogance of the standards. The standards support the largely unproven assumption that all of this mathematical content is necessary to be a productive citizen in our democracy. The standards also support the unproven assumption that mathematical reasoning is similar to or transfers to other types of logical reasoning. Combine both assertions with the call for increased high school requirements for mathematics (and the largely impractical word problems that will be generated for study), and the standards will produce exactly the opposite of what is intended: instead of mathematics for the everyman, there will exist mathematics for the elite and math anxiety for everyone else.

The standards continually emphasize that what they outline is needed for all students to become "productive citizens in the twenty-first century." To the credit of the standards' authors they do see the problem in the inequality of mathematical knowledge:

> If all students do not have the opportunity to learn this mathematics, we face the danger of creating an intellectual elite and a polarized society. The image of a society in which a few have the mathematical knowledge needed for the control of economic and scientific development is not consistent either with the values of a just democratic system or with its economic needs.[14]

In one sense this argument is paradoxical: by arguing that understanding mathematics increases the ability to adapt to change, to reason logically, and to increase self-confidence it is essentially being posited that those who have such understanding are superior. Once again, this is the claim that mathematical knowledge is primary. Very little is mentioned in the standards about the importance of other disciplines. By not seriously confronting how mathematics is linked to other areas and to the practical, and complex, problems of the everyday citizen, the standards may create an elite by default: those who do well on these "word problems" may not only see themselves as superior but may feel entitled to control whatever they can.

The Argument Against a National Examination

Both national standards and a national assessment in mathematics were supported during the George Bush administration and have continued to receive encouragement during President Clinton's tenure. In 1989, during the President's Education Summit, then President Bush and the state governors adopted six National Education Goals, designed for all Americans to reach by the year 2000. We have already discussed the fourth goal, which advocates that the United States be number one in the world in mathematics and science (see chapter 5). Goal number three, which was concerned with student achievement and citizenship, directly advocated a particular shape for the curriculum:

> By the year 2000, American students will leave grades four, eight, and twelve having demonstrated competency in challenging subject matter including English, mathematics, science, history, and geography; and every school in America will ensure that all students learn to use their minds well, so they may be prepared for responsible citizenship, further learning, and productive employment in our modern economy.

As a consequence of adopting these National Education Goals, Congress created the National Education Goals Panel, whose job is to report on progress toward these goals, until the year 2000. After its first year, this Goals Panel suggested that, to adequately measure goals three and four, national education standards would have to be defined. Consequently, The National Council on Education Standards and Testing was created to advise Congress on the feasibility of both national standards and a national test. This task force was co-chaired by Governor Carroll Campbell of South Carolina and Governor Roy Romer of Colorado.

As might have been anticipated, the report of this task force, *Raising Standards for American Education,*[15] released in January 1992, advised that national standards and a national system of assessments were both desirable and feasible. Without high national standards, the authors of the report argued, the educational system is somehow disappointing its customers: "In the absence of well-defined and demanding standards, education in the United States has gravitated toward de facto national minimum expectations. Except for students who are planning to attend selective four-year colleges, current education standards focus on low-level reading and arithmetic skills and on small amounts of factual material in other content areas. Consumers of education in this country have settled for far less than they should and for far less than do their counterparts in other developed nations."[16]

What would national standards and assessments do for America? The Council advocated that high national standards and assessments could accomplish three purposes. First, high standards could produce high expectations for all students, particularly those who traditionally have had less academic success. These standards would stimulate the educational system to provide resources so that all American students could reach these goals. Second, high standards would enhance our "civic culture" and help all students "acquire the necessary knowledge, skills, and shared values to deepen and renew our civic culture and to enable all citizens to participate more effectively in the political processes of democracy."[17] Finally, high standards would enhance America's economic competitiveness.

Once again, in the case of the third purpose, the success of business and the education of our students are yoked together, as we saw earlier in the argument for being number one in the world in math and science. This report is even more straight-forward: "The relative deficiency in America's human capital contributes to the inability of many firms in the United States to compete internationally. Low skill levels may also be impeding American business from shifting to newer, more efficient methods of production that require greater responsibility and skill on the part of front-line workers."[18]

The rationale for these high standards clearly parallels many of the arguments that had been advanced by the National Council of Teachers for Mathematics (NCTM) to support their version of a revised mathematics curriculum. Since the NCTM standards had been released in early 1989, the Council on Standards and Testing cite the NCTM document often in their report as a model for what the forthcoming national standards might encompass.

Two of the arguments that were advanced above against national standards in mathematics apply equally well to national standards in other subject matter. First, the argument for mathematics came from superiority, i.e., that mathematics is so essential for all facets of everyday life. Second, mathematics educators will not stand for any of their material to be deleted from the curriculum; in fact, their standards propose much more content that will have to be studied by students at all levels.

Since the release of the report on the desirability of national standards, other disciplines have been quick to borrow these two arguments from the mathematicians and use them to support their own national standards. National Education Goal number three had given primacy to five subject areas: mathematics, English, science, history, and geography. Other disenfranchised groups quickly argued to be on this national bandwagon: among these groups, so far, are the arts, civics, economics, foreign languages, physical education, and social studies.

Each discipline has established a task group, collected its experts, and begun assembling national standards in their areas that *all* Americans will need to know and to be able to use or else they won't be productive citizens. If each discipline imitates mathematics and advocates three to four years of study in high school for their subject matter, then the average high school student, the future American citizen, should be prepared to take eleven courses a day (this figure, of course, only represents the eleven disciplines which have come forward so far)!

For instance, early drafts of the geography standards suggest that all Americans will need to be aware of the following: nature and distribution of ecosystems on Earth's surface; nature, distribution, and migration of human populations; nature and complexity of Earth's cultural mosaics; and Earth's physical and human systems, their connections and interactions. The music standards are equally interesting. They suggest that students will need to be able to improvise music in a variety of styles, including the ability to improvise "music that demonstrates originality and imagination in choices of tempo, rhythm, timbre, dynamics, and sound sources for expressive purposes."[19] Furthermore, students should be able to read musical notation, and should be able to "sing or play music from notation using rhythmic values of whole, half, quarter, eighth, sixteenth, and dotted rhythms in 2/4, 3/4, 4/4, 6/8, 3/8, and 2/2 meters."[20]

No discipline seems to have heard of the idea of cooperation or compromise. Most of the standards suggest the importance of studying the interaction of their material with other disciplines, as did the mathematicians. But actually confronting how this interaction is to be accomplished in a realistic, time-limited setting is an extremely difficult task; attempts to construct a more modest curriculum will be discussed in the next section.

In addition to proclaiming the desirability of national standards, the Council on Testing and Standards recommended the feasibility of a national system of assessments. This system of assessments would measure individual student progress and would also consist of large-scale assessments, to report on national progress toward the education goals. The council did suggest that the system of assessments should not consist of a single test, but be composed of multiple methods of measuring progress. In this regard, it parallels the recommendations of the NCTM, which also suggested multiple ways of measuring student progress. To their credit, the council members also suggested that these new assessments be aligned with the new standards: "If we believe that we should be encouraging the development of extended mathematical reasoning and problem-solving, assessment must include complex tasks and ample time for thought."[21]

I am very pessimistic that a national examination in mathematics would have any of the idealistic features advocated by this report. I believe any national examination that is constructed would consist of very time-limited mathematical exercises, which not only would emphasize the sophistical conceptions of the mathematical triumvirate—algebra, geometry, and trigonometry—but would also consist of more impractical word problems. On what do I base this belief?

For over twenty years, this country has had a de facto national examination in mathematics. At first glance, this examination might seem to be the SAT, which, as we have seen from earlier chapters, has been used as a barometer of student progress. The federal government, however, has funded directly a National Assessment of Educational Progress (NAEP), whose mission is to measure student progress in five subject areas: mathematics, science, history, geography, and English skills. (It is no coincidence that the five subject areas listed in National Education Goal number three are the disciplines for which current NAEP exams exist. Furthermore, as I outline the specifics of the NAEP procedure, it should come as no surprise that the Educational Testing Service, which as a not for profit corporation administers the SAT, also has the federal contract to conduct the NAEP studies.)

The NAEP measures student progress in these five subject areas approximately every two years. Over the past twenty years, students in grades four, eight, and twelve have been tested regularly in mathematics. Since the early 1970s, overall averages on the NAEP mathematics test have changed only slightly: during certain periods the scores have declined and during other periods the scores have risen moderately. On more challenging, complex mathematical material, however, American students, especially the seventeen-year-olds who are about to graduate from high school, have displayed mediocre results for close to twenty years.

As one author has suggested, commenting on the results of the 1986 NAEP mathematics assessment:

> Performance by high-school students was even more unsettling [than those of students in lower grades]. Although students graduating from high school seem to be able to add, subtract, multiply, and divide, this level of achievement is hardly in the spirit of the country's goal, which is grounded in competency with challenging subject matter. Only half the seventeen-year-olds assessed in 1986 demonstrated a grasp of even moderately challenging mathematical procedures and reasoning (i.e., deci-

mals, fractions, and percents; simple equations), and only 6 percent reached the highest level of proficiency defined—a level characterized by a high rate of success on questions measuring multi-step problem solving and algebra.[22]

This inability to deal with challenging subject matter has dire consequences for the economy, as this particular NAEP report also emphasized our inability to compete internationally. Furthermore, without these mathematical and reasoning skills, students are at a disadvantage

to solve problems related to the future health of our country and planet. Our daily lives are influenced by the rise and fall of the stock market, the size of the government budget and budget deficit, the balance of trade with other countries, and international monetary policy. Debates are carried on about world-wide environmental issues, including depletion of the ozone layer, acid rain, and global deforestation. Unfortunately, there is growing evidence that the typical American high-school student does not know enough to appreciate what these issues or debates are all about, let alone to participate creatively and effectively in making decisions about the salient issues.[23]

Are these NAEP examinations really providing us evidence to make such judgments about the intellectual skills of students and the future of our country? Should what NAEP measures become our actual national examination, required of all students? Let's examine the NAEP mathematics examination a little more closely.

The NAEP mathematics test is scored on a scale ranging from 200 to 350, with four discrete levels. A summary of these four levels and the abilities they measure is as follows:

Level 350: Reasoning and Problem Solving Involving Geometric Relationships, Algebra, and Functions;

Level 300: Reasoning and Problem Solving Involving Fractions, Decimals, Percents, Elementary Concepts in Geometry, Statistics, and Algebra;

Level 250: Multiplication and Division, Simple Measurement, and Two-Step Problem Solving;

Level 200: Addition and Subtraction, and Simple Problem Solving with Whole Numbers.

The top level of 350 is most crucial; these problems represent the ability to tackle challenging subject matter and to be prepared for the workplace of the future. Student performance at this level, however, has been quite discouraging. In 1978, only 7 percent of seventeen-year-olds made it to the top level; in 1982, the figure was 6 percent; in 1986 it rose to 7 percent; and in 1990 it reached 7 percent once again. If this report is correct, practically no one who has graduated from high school in the last twenty years has been prepared for a challenging job. It's amazing that the country has continued running at all, given these gloomy numbers.

Just what types of problems are our seventeen-year-olds unable to solve? On the next page are eight problems from the NAEP Level 350. See how many you can do.[24]

Are you adequately prepared for the job you presently have? Unless you worked a majority of these problems correctly, NAEP might argue that you are underprepared for the business and economic demands of the twenty-first century. Furthermore, you are not prepared to partake in discussions of complex social issues, such as the budget deficit and the acid rain controversy.

I contend, however, that not only is the NAEP examination

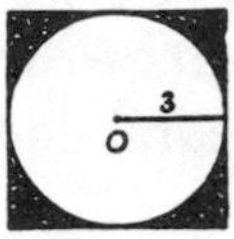

1. In the figure above, a circle with center O and radius of length 3 is inscribed in a square. What is the area of the shaded region?

A. 3.86

B. 7.73

C. 28.27

D. 32.86

E. 36.00

2. Suppose $4r = 3s = 10t$, where r, s, and t are positive integers. What is the sum of the least values of r, s, and t for which this equality is true?

A. 7

B. 17

C. 41

D. 82

E. 120

3. In the xy-plane, a line parallel to the x-axis intersects the y-axis at the point (0, 4). This line also intersects a circle in two points. The circle has a radius of 5 and its center is at the origin. What are the coordinates of the two points of intersection?

A. (1, 2) and (2, 1)

B. (2, 1) and (2, –1)

C. (3, 4) and (3, –4)

D. (3, 4) and (–3, 4)

E. (5, 0) and (–5, 0)

$x^2 + y^2 = 25$

$x^2 = \pm 9 \; x = \pm\sqrt{3}\sqrt{3}$

4. For what value of x is $8^{12} = 16^x$?

A. 3

B. 4

C. 8

D. 9

E. 12

$(2^3)^{12} = 2^{36}$

$16 (2^x)^y = 2^{4x}$

5.

60

The area of rectangle $BCDE$ shown above is 60 square inches. If the length of AE is 10 inches and the length of ED is 15 inches, what is the area of trapezoid $ABCD$, in square inches?

Answer: _____ 80 inches

6. If $f(x) = 4x^2 - 7x + 5.7$, what is the value of $f(3.5)$?

30.2

Answer: _____

Figure 3

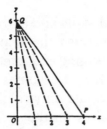

7. In the figure above, point Q is fixed and point P starts at 4 and moves left along the x-axis. As P moves left along the x-axis toward O, the area of ΔPOQ changes.

    Use the information given to complete the table below to show how the area of ΔPOQ changes as P goes from the position shown to the origin O.

x - coordinate of P	Area of ΔPOQ
4	12
3	9
2	6
1	3
0	0

8. The figure above shows the graph of y = f(x). Which of the following could describe the graph of y = |f(x)|?

A.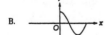

B.

C.

D.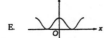

E.

Figure 3 (con't)

a perpetuation of archaic mathematical skills and extremely short-sighted word problems, it also exaggerates the impact of these skills on the ability to solve really challenging and interesting problems. To advance NAEP to the point that it becomes a mandated national examination for all students would only reinforce the impracticality of the mathematics curriculum for the average citizen.

Just how can students be prepared to handle complex problems? In the next section, I will examine some more modest attempts to fashion a realistic curriculum and examination system. For this system to work, there can no longer be separate standards in mathematics, a separate examination in mathematics, nor even separate courses in mathematics, especially at the high school level.

Solving Complex Problems

How much mathematics is necessary? If there are to be no standards and no national examination, just what exactly is going to be studied? If there are not even to be separate courses in mathematics, what will students spend their time doing, especially in high school? In formulating answers to these questions, it helps to examine one of the arguments that led to the revamping of the mathematics curriculum. Part of the rationale for change sprang from an economic motive: America is falling behind the world economically so therefore we need to train our students to cope with the jobs of the future.

Just what are these jobs of the future? Robert Reich, in his book *The Work of Nations*,[25] discusses changes in the connection between corporations and the national identity and outlines the three types of jobs he thinks will be valued in the coming years. Reich notes that the concept of the corporation of the 1950s—

as GM goes so goes America—has changed dramatically. No longer do most modern corporations control the entire production process for their product. No longer are products even "national." The new American car that you may be driving was probably assembled from parts manufactured around the world. Most corporations today are international in scope and compete by blending products and services from all over the globe. Since products and corporations are no longer strictly national, all that remains rooted within a country are its people. Thus to survive in this world of the future, each nation must develop its main assets: its citizens' skills and insights.

To this point Reich's argument is reasonably similar to that used by the mathematicians to justify their standards. In outlining what he calls the three jobs of the future, however, Reich reconceptualizes just what "skills and insights" are to be valued. For instance, one of these jobs will be routine production services. These jobs involve simple repetitive tasks of any nature, whether it be assembly-line production, routine checks of computer programs, or order processing. Workers in these jobs must be able to read, perform simple calculations, and take direction. The problem, as Reich sees it, is that these jobs are being lost to the cheaper and more competitive world marketplace and will not return.

The second category of jobs, in-person services, also entails simple and repetitive tasks. These jobs involve personal contact with other people—from restaurant servers, to janitors, health-care aides, flight attendants, and security guards. Persons in these occupations tend to have good interpersonal skills and are reliable. As Reich notes, these jobs do not leave the country, as do routine production services. On the other hand, these jobs do not make us competitive; they often arise from the success of other endeavors.

The jobs that will make us competitive are those involving

symbolic-analytic services. These services comprise three activities: *problem-solving skills,* the ability to put things together in unique ways; *problem-identifying skills,* the ability to identify new problems and possibilities for the unique product that has been devised by the problem solvers; and, finally, *strategic brokers* who have the savvy to connect skillfully problem solvers and problem identifiers. Symbolic-analytic services deal with the manipulation of symbols—whether it be data, words, visual or oral media—and can be traded worldwide. Specific occupations that utilize these skills range from investment bankers and lawyers, to software engineers and research scientists, to systems analysts and strategic planners, to writers, journalists, film producers, and publishers.

So how should the future symbolic analyst be educated? Reich identifies four basic skills that are needed: abstraction, system thinking, experimentation, and collaboration. *Abstraction* is the ability to manipulate and assimilate huge masses of information into some new pattern. These new patterns often create new realities which engender new solutions to problems; witness the work of the innovative scientist, lawyer, or screenwriter. Traditional education does not train students for this type of abstraction. "Rather than construct meanings for themselves, meanings are imposed upon them. What is to be learned is prepackaged into lesson plans, lectures, and textbooks. Reality has already been simplified; the obedient student has only to commit it to memory."[26] The training of the symbolic analyst is different:

> The student is taught to get *behind* the data—to ask why certain facts have been selected, why they are assumed to be important, how they were deduced, and how they might be contradicted. The student learns to examine reality from many angles, in different lights, and thus to visualize new possibilities and

choices. The symbolic-analytic mind is trained to be skeptical, curious, and creative.[27]

System thinking accentuates the ability to see how certain problems are interconnected with other problems and causes, that, for instance, a computerized workstation for the home may solve problems associated with rush-hour traffic. "Rather than teach students how to solve a problem that is presented to them, they are taught to examine why the problem arises and how it is connected to other problems."[28] Most traditional education, as Reich comments, does not emphasize interconnectedness but compartmentalization, giving students "courses" in separate disciplines, such as mathematics, with very little attempt to show how one discipline is connected to another. Experimentation increases the capacity for creativity and for new solutions. But it also exposes the personal and emotional concomitants of trying to be creative—the risks that must be taken, the frustration and disappointment that come from false starts and failed attempts, and even the fear that could arise from venturing into the unknown. Once again traditional education falls short: "The tour through history or geography or science typically has a fixed route, beginning at the start of the textbook or the series of lectures and ending at its conclusion. Students have almost no opportunity to explore the terrain for themselves. Self-guided exploration is, after all, an inefficient means of covering ground that 'must' be covered."[29]

Finally, the *ability to collaborate* is important for the future symbolic analyst. Most modern problems are so huge and complex that they are approached by teams of workers. Thus the abilities to communicate ideas, share concepts, and reach consensus are very valuable. As Reich comments, in this country, most symbolic analysts live in specialized zones which nurture and support their increasing skill and innovation—just consider California's Silicon

Valley for computer hardware and software; Los Angeles for music and film; New York and Chicago for global finance; and a host of other specialized areas. These zones have become "dynamic learning communities"[30] and the future symbolic analyst must learn how to thrive in this competitive culture.

Reich notes that the type of learning needed to foster abstraction, systems thinking, experimentation, and collaboration already occurs in some of the nation's best high schools, universities, and graduate schools. What if this approach to learning were more widespread? What if every high school student had the opportunity to master these skills? What needs to be done to break from traditional compartments of knowledge—such as the high school algebra or geometry course—in order to encourage this type of meaningful, creative, complex thinking?

Theodore Sizer has the best practical vision of what high school should be. Since 1984, Sizer, his colleagues at Brown University, and an association of school personnel have promoted the Coalition of Essential Schools. Their vision is to transform schooling and especially the high school experience such that "schools focus on helping adolescents to use their minds well."[31] As Sizer argues, the essence of the high school experience should be to help students engage thoughtful and complex problems and to make school more meaningful to the lives of its students.

The centerpiece of Sizer's reforms is focused on what he calls the *exhibition*. Each high school should make it very clear to its students what outcomes are valued and exactly what it expects the student to be able to do or perform by the time he or she graduates high school. These displays of knowledge and reasoning would then be exhibited, for other teachers, students, and parents to see. No longer would students necessarily receive report cards with grades in courses. Instead, they would be demonstrating their abilities to do more meaningful and realistic tasks.

Let's look at some sample exhibitions to see what Sizer has in mind. Here is one that is entitled Form 1040:[32]

> Your group of five classmates is to complete accurately the federal Internal Revenue Service Form 1040 for each of five families. Each member of your group will prepare the 1040 for one of the families. You may work in concert, helping one another. "Your" particular family's form must be completed by you personally, however.
>
> Attached are complete financial records for the family assigned to you, including the return filed by that family last year. In addition, you will find a blank copy of the current 1040, including schedules, and explanatory material provided by the Internal Revenue Service.
>
> You will have a month to complete this work. Your result will be "audited" by an outside expert and one of your classmates after you turn it in. You will have to explain the financial situation of "your" family and to defend the 1040 return for it which you have presented.
>
> Each of you will serve as a "co-auditor" on the return filed by a student from another group. You will be asked to comment on that return.
>
> Good luck. Getting your tax amount wrong—or the tax for any of the five families in your group—could end you in legal soup!

This exhibition embodies many elements which contrast it strongly with the "exercises" or "word problems" that are often encountered in many classes. First, the task is authentic and important. In fact, for many adults (probably including some of the students' parents), this whole process might still be confusing, intimidating, and scary. Second, it should appeal to students, in that it addresses both a general issue in society and a problem that will become personal to them. If the families who are helped have drastically different incomes, it should address the concerns of money, power, and some of the inequities that arise from these issues.

Third, the completion of this task exposes students to knowledge in several disciplines: mathematics, microeconomics, politics, ethics, and political history, for example. It should also provoke questions about equality: who, for instance, is entitled to "tax breaks" and for what reasons? Fourth, it focuses on both individual and collaborative work, skills which are necessary in the workplace. Fifth, the task will take time and energy; it thus reinforces persistence, organization, accuracy, and careful reflection on what the student is trying to accomplish. Finally, it tells students that their work is being taken seriously. By presenting the results of this exhibition, and having to defend what has been done, students can sense that what they do is valued and respected by adults around them.

Let's look at another example:[33]

You have decided to buy a used car. You must choose between a 1988 Ford Tempo, priced at $5,800, and a 1988 Toyota Corolla, priced at $6,700. Both cars have had one owner and been driven about 36,000 miles; both have four doors and mediocre but working stereo systems. A mechanic friend of yours has verified that they are both in good condition.

You have saved $1,100. Since you are too young to qualify for a bank loan, your parents will lend you the remaining purchase cost at the terms available from a bank. They will also add your car to their insurance policy, but you must pay the additional premium. You have a part-time job, at which you work eighteen hours a week and for which you are paid 15 cents per hour over the minimum wage. Your employer likes your work and has told you that you will get a raise of 12 cents an hour every six months, beginning three months from now. For ten weeks in the summer, you may work forty hours a week, less any vacation time you want to take.

Determine the cost of buying and operating each of the cars over four years. Indicate which car you will buy, describing your reasoning and showing your calculations. Assuming that you must spend on entertainment as well as on your car, explain in detail the implications of purchasing whichever auto you decide on.

Once again, this task cuts across several disciplines and draws on knowledge (which the student must seek out) from areas such as mathematics, civics, economics, and even labor relations. This task is also authentic and very likely one that the student is facing or could be facing. This task also asks for realistic decisions and asks participating students to defend their decisions.

It must be understood that the solution of these problems, and the students' resulting exhibition of the results, are not to occur *in addition to* traditional courses; instead, these exhibitions are the courses, so to speak. As teachers and students decide on what will constitute meaningful exhibitions of a student's performance, then these performances become the focus of instruction and problem solving in the high school. Students spend their time working on these problems and the resources of the high school are marshalled to help them become proficient in the skills and abilities that are desired. If courses existed at all, they would only be secondary or supportive of the students' completion of their exhibitions.

Exhibitions can also be centered on the complex world of social and international issues and the ability of citizens to discuss these concerns. Consider the following problem of my own creation:

What effect might the breakup of the former Soviet Union
have on citizens in the United States?

There are many approaches to this problem and different types of evidence that students might assemble to bolster their arguments and opinions. Are there mathematical or economic models which address the dissolution of countries? Is there historical evidence on the breakup of empires and does it apply in this situation? Should we examine the attitudes of Americans toward Russia and communism and how these might be changing? Should our political policies and foreign aid change? There are certainly many more questions to be asked, each requiring students to think for themselves and to assemble and evaluate information that pertains to a complex problem.

What would a high school look like if it were entirely focused on exhibitions? Sizer outlines several characteristics of a good school. A good school is a place that respects students and does not view them as irresponsible. As Sizer comments, "Most schools infantilize and pamper their students. They are not to be trusted: all will need hall passes to go somewhere when classes are in session."[34] Students are usually not given responsibilities for cleaning up or helping to maintain the school; thus, it is more difficult for them to begin to take responsibility for their own lives.

A good school also helps students learn by doing and using, not just by memorizing. To facilitate learning by doing, the student must be engaged in problems and activities that are interesting and meaningful. Furthermore, learning by doing will involve a great deal of talking: "Good schools are suffused with talk, with all sorts of constructive conversation."[35] Learning is not most effective with a teacher lecturing and students quietly sitting in their seats.

A good school also cultivates the habit of thoughtfulness in its students. In adult society, we often admire people's habits more than the factual knowledge they can recite. As Sizer comments:

> We admire the person who can be counted on to think before acting, to weigh matters before dealing with them. . . . We value the individual who can cope with something new. . . . We admire the person who does not panic or disappear when something new or threatening comes into the community. . . . We respect the individual who regularly finds something new even in the familiar, who looks for reasons to laugh at the absurdities in life, and who enjoys sharing that laughter.[36]

A good school cultivates intellectual habits in its students. The habit of perspective helps students sort the important from the unimportant in interpreting arguments and making decisions. The habit of analysis teaches students how to use various tools—mathematical, logical, artistic, and aesthetic—in evaluating the evidence that is presented in certain arguments. The habit of imagination helps student to consider new possibilities and to value their own creativity. The habit of empathy allows students to understand other persons' points of view. The habit of communication helps students learn to communicate their own ideas and feelings, and the various means of communications, as well as helping them become good listeners. The habit of commitment stresses the importance of persistence and the ability to act when action is called for. The habit of humility notes the limitations that each person has and underlines what one knows and does not know. Finally, the habit of joy reinforces students finding excitement and interest in many things in life.[37]

National standards and a national examination, in any subject matter, are almost beside the point when it comes to cultivating the real habits of problem solving and reflection, as Sizer defines them. These same standards and examinations are also naively simplistic solutions to understanding how the complicated process of thinking can be taught. As Sizer notes:

> Much contemporary discussion of standards follows a seductively simple argument. National and state policymakers must decide on the standards to be met. A syllabus will be issued to help schools aim at these standards. Tests that measure achievement of those standards will be administered to all kids. Those who pass will get diplomas. Schools will be compared, and those which have lots of failing students will be humiliated into reform or be taken over by higher authorities.[38]

This argument has a number of flaws, many of which have already been discussed. There is the problem of who shall set the standards; as we have already seen, policymakers in each discipline are extremely arrogant in assuming that almost everything in their discipline is important for all students to know. Standards also suggest that there is very little disagreement among scholars, within and across disciplines, as to what is valuable in their field. Furthermore, intellectual freedom becomes implicitly curtailed: for, if students want to question the standards, what recourse will they have?

A single or even multiple system of national examinations shares some of the same flaws. As I have tried to demonstrate, with numerous examples here and in other chapters, mathematicians cannot seem to devise anything but the most simplistic and arcane measurements of their discipline. A national examination would probably perpetuate this obtuse knowledge, with the side effect of making generations of students feel incompetent in doing mathematics and reinforcing their belief that mathematics does not really apply to anything outside of school. Finally, narrow, time-limited examinations are not reflective of the habits and skills that need to be cultivated in our students.

The schools envisioned by Sizer and his colleagues do not relinquish the notions of substance and standards in the curriculum, nor do they dismiss any type of examination. By focusing on exhibitions, however, they face head on the difficulties of students becoming engaged in the learning process, attempting to solve complex problems, and understanding that schooling matters in the development of self and in the world outside school. Exhibitions dispel the notion that tests are only for classrooms and only examine secret or guarded knowledge. Exhibitions reinforce the notion that students are expected to master knowledge deeply and to be able to use it in public situations that call for dialogue and follow-up. Exhibitions also suggest that "standards" evolve, as students and teachers attempt to work through complex problems.[39]

Avoiding Number Nonsense

How are we to avoid number nonsense? First, claims by mathematicians and math educators that everyone should study mathematics, especially the abstract courses offered in high school, should be examined very closely. The question of how much mathematics is really necessary for life should be raised constantly by students, parents, and teachers.

Second, claims that studying mathematics helps someone develop their logical reasoning abilities should not be taken any more seriously than claims that studying other rigorous disciplines helps to develop these faculties. A person is not illogical or stupid merely from the inability to solve the types of problems that some mathematicians find aesthetically appealing and interesting.

Third, any tests that are mandated in a subject like mathematics should first be taken by state legislators. If mathematics is so important for success in life, and everybody needs to know this material, then let the "standards" be set by the policymakers. Let mathematics tests be normed on these policymakers, and national and state legislators, and reform advocates. Let legislators publicly display how much of this knowledge they possess, possibly by posting their scores on the NAEP exam.

Finally, opposition to national standards and a national examination should help to refocus the debate within educational circles on the learning process. It is a difficult matter to learn to use material in complex situations, and the imposition of standards and examinations will not guarantee that this type of learning takes place. Much less mathematics is needed for success in our society than most mathematicians think. What is needed are more honest approaches to the challenging demands that a democratic society presents to its citizens and the realistic problems that might be faced in a lifetime. If some mathematics is needed to master these challenges, then this discipline should be studied. If very little mathematics is needed for the complex problems of a democratic society, then we should have the courage to drop this discipline and have education focus instead on what is more important.

Notes

1. National Council of Teachers of Mathematics, *Curriculum and Evaluation Standards for School Mathematics* (Reston, Va., 1989).

2. Ibid., p. v.

3. Edward B. Fiske, *Smart Schools, Smart Kids: Why Do Some Schools Work?* (New York: Simon and Schuster, 1991), p. 25.

4. National Council of Teachers of Mathematics, *Curriculum and Evaluation Standards for School Mathematics*, p. 3.

5. Ibid., p. 5.

6. Ibid.

7. Ibid.

8. Ibid., p. 6.

9. Ibid., p. 139.

10. Ibid., p. 137.

11. Ibid., p. 132.

12. Michael W. Apple, "Do The Standards Go Far Enough? Power, Policy, and Practice in Mathematics Education," *Journal for Research in Mathematics Education* 23, no. 5 (1992): 424.

13. National Council of Teachers of Mathematics, *Curriculum and Evaluation Standards for School Mathematics*, p. 124.

14. Ibid., p. 9.

15. The National Council on Education Standards and Testing, *Raising Standards for American Education* (Washington, D.C., 1992).

16. Ibid., p. 3.

17. Ibid., p. 10.

18. Ibid., p. 11.

19. Debra Viadero, "Geography Educators Release First Draft of Curriculum Standards," *Education Week*, August 4, 1993, p. 5.

20. Debra Viadero, "Draft Standards for Arts Education Stress Both Knowledge, Performance," *Education Week*, August 4, 1993, p. 5.

21. National Council on Education Standards and Testing, *Raising Standards*, pp. 1–2, 3.

22. Ina V. S. Mullis, Eugene H. Owen, and Gary W. Phillips, *America's Challenge: Accelerating Academic Achievement. A Summary of Findings from 20 Years of NAEP* (Princeton, N.J.: Educational Testing Service, 1990), p. 20.

23. Ibid., p. 7.

24. Problems taken from Ina V. S. Mullis, John A. Dossey, Eugene H. Owen, and Gary W. Phillips, *NAEP 1992: Mathematics Report Card for the Nation and the States* (Princeton, N.J.: Educational Testing

Service, 1993), pp. 60–63, 257–61.

25. Robert B. Reich, *The Work of Nations* (New York: Vintage Books, 1992).

26. Ibid., pp. 229–30.

27. Ibid., p. 230.

28. Ibid., p. 231.

29. Ibid., p. 232.

30. Ibid., p. 240.

31. Theodore R. Sizer, *Horace's School: Redesigning the American High School* (Boston: Houghton Mifflin, 1992), p. 207.

32. Ibid., p. 48.

33. Ibid., p. 100.

34. Ibid., p. 59.

35. Ibid., p. 89.

36. Ibid., pp. 69–70.

37. Ibid., pp. 73–74.

38. Ibid., p. 110.

39. Joseph P. McDonald, Bethany Rogers, and Theodore R. Sizer, "Standards and School Reform: Asking the Essential Questions," *Studies on Exhibitions*, no. 8 (1993).